AF479321

Digital MOTHERHOOD

PARENTING WITH TECH IN THE EARLY YEARS

Amanda Strawhacker, Ph.D.

Amanda Sullivan, Ph.D.

Text copyright © 2026 by Amanda Strawhacker, PhD and Amanda Sullivan, PhD
All rights reserved. Printed in the United States of America
Published by Motina Books, LLC, Bailey, CO
www.MotinaBooks.com

Library of Congress Cataloguing-in-Publication Data
Names: Strawhacker, Amanda and Sullivan, Amanda
Title: Digital Motherhood
Description: First Edition. | Bailey, CO Motina Books, 2026

Identifiers:
LCCN: 2026930129

ISBN-13: 979-8-88784-071-0 (hardcover)
ISBN-13: 979-8-88784-069-7 (ebook)
ISBN-13: 979-8-88784-070-3 (paperback)

Subjects:
FAMILY & RELATIONSHIPS / Parenting / General
FAMILY & RELATIONSHIPS / Life Stages / Infants & Toddlers
TECHNOLOGY & ENGINEERING / General

Dedication

FOREWORD

When I was growing up, my brothers and I gathered around the one television in our house. It was in the den, and it was a black-and-white TV. By the time my own children came along, high-def television, cable, and some on-demand services were the norm, we had more than one TV in the house, and, of course, everything was in color. I distinctly recall twirling the plastic cord of the old black rotary phone in the basement of my childhood home as I spoke with friends (hopefully out of earshot of my brothers or parents); my own children grew up as smartphones increasingly became a must-have item and they could speak or text with their friends, anywhere. I remember the first computer I ever saw was the mainframe in the subterranean room under my graduate school at the University of Pennsylvania: it was huge and went across an entire wall; different colored lights flashed in patterns I could not understand. We had to feed punch cards into it to process data, and this took hours to do. My own children grew up with ever-smaller, sleeker, faster laptops that could perform numerous complex functions within seconds.

I mention this not just to show that within the span of a relatively few short decades technology has changed tremendously and irrevocably, but to point out that parents *always* grow up in a world of different technologies than their own children do. It is parents who often need to get up to speed, to adapt, to adjust.

Having children and watching them engage with media often makes you think, "it wasn't like this when I was a kid!" The truth is, it wasn't. But in some ways, it also was: what a lot of the research about children and media use shows us, more than anything, is that the more things change, the more consistent patterns emerge. Knowing this, recognizing those patterns, can help us to help our children sort through an ever-shifting media landscape.

Children today are growing up with a dazzling array of media. We all grew up in a media-saturated world. But for children today, media is wall to wall, around them all the time, available to them across increasing numbers of platforms. It's an important part of how they take in, define and process knowledge, and shapes, in powerful ways, how they see the world.

Today's children don't know a time without smartphones and the internet. They are growing up in an age where information is always, literally, just a tap away, and where entertainment comes in shorter and shorter segments, available to them any time, any place. Children of the 21st century have social lives and friendships created and fostered by their media use. Many of their communication patterns with peers revolve around shared media content. It is no surprise, then, that they spend a considerable amount of their time every single day with technology. Most of the recent research shows that children today spend more time engaging with media of one sort or another than they spend in school; they spend more time with media than with any other activity, except for sleeping. It's not a surprise that across all age groups, and across all demographic groups, children's screen time on different platforms went up during the pandemic; what's a bit more surprising, perhaps, is that it doesn't seem to have gone back down.

Perhaps the single biggest change in children's media use over the past decade has been the move toward mobile. More kids have access to mobile devices, at younger and younger ages, and this is true across every demographic group. This has profound implications for all kinds of ways that kids use media, what they see and hear and what they do. It also often means that we, as parents, don't know what our kids are viewing, listening to, or playing. Children are using media in increasingly privatized ways, and we just don't know what the long-term implications of this might be.

As someone who does research on and teaches about children and media, and as the mother of four children, I'm often asked for advice about this brave new world of technology. How much screen time is too much screen time? When should you give in and get your child their first phone?

How can you help your child navigate the often confusing and ever-changing world of social media, especially if it's a world that you don't understand too well, yourself? I've always tried to be more proactive than prescriptive. I've tried to balance what I know from the research with what I know from the realities of being a mom. And I've always tried to center my belief that rather than restrict kids' access to media, which is sometimes frustrating and often, futile, it's better to try to do what you can to make your children media literate so that they can make good choices, themselves.

Amanda Strawhacker and Amanda Sullivan's book is an important resource for today's parents because it's filled with good, practical advice from people who grew up in a generation that had a lot of media at its disposal and are now parents themselves. But it's also an important resource because their advice is grounded in up-to-the-minute research. Having had the great pleasure of teaching both Amandas when they were graduate students, I know them both as people who care deeply about making academic research more accessible, getting research into the hands of the people who can most benefit from it. *Digital Motherhood: Parenting with Tech in the Early Years* is a wonderful blend of evidence-based advice and honest, practical mom experience. This is the kind of go-to resource we really need at a time when media platforms change almost daily and as our children grow up in a world that's ever-more media saturated.

Julie Dobrow, PhD
Tufts University

INTRODUCTION:

NEW TECHNOLOGY AND CONCERNS FOR THE MODERN MAMA

As we write the introduction to this book, we are currently moms to a curious five-year-old boy and a very energetic two-year-old girl (Amanda Sullivan), and a newborn baby girl (Amanda Strawhacker). We have found the rare quiet moment during their afternoon naps or between nursing to begin writing for the very first time today. There were many moments over the past few months when one of us had a flash of an idea or a stimulating conversation and found ourselves with the urge to write. But, we had to settle for a quick note texted to our co-author here, or an outline on our phone's note-taking app there, because pulling out the laptop just isn't always practical when chasing around young children. Moreover, we strive to be deliberate about when and how we use personal technology around our kids (keyword here being *strive,* meaning we don't get it perfect every time either!).

In fact, from the moment we began our motherhood journeys, we have been, and continue to be, confronted by technology and digital media almost every moment of every day. From apps that told us what fruit our unborn babies were the size of while we were pregnant (at 36 weeks they're the size of a pineapple, if you were wondering) to the television shows Amanda Sullivan's son is slowly becoming obsessed with (Octonauts, anyone?), we can't escape the digital-media-rich environment in which we are increasingly steeped. Like other moms, we have to make decisions each day about the "technology diet" that our families will consume. And we know that our own cell phone, tablet, and laptop use—whether for work, writing, or personal use– is just as important for us to consider as our kids' technology use.

Perhaps unlike other moms, we are also both researchers and educators in the field of educational technology and child development. Independently and collaboratively, we have worked in this field for over a decade before having our own children, earning our Ph.D.s by studying the impact of new technologies and media on young children at the Developmental Technologies ("DevTech") Research Group at Tufts University. During our time at DevTech, we researched the many positive developmental benefits technology offers young children and some of the negative aspects we want our families to avoid. Does that give us an advantage and make it any easier on a daily basis? Nope. But it *does* give us the research knowledge and understanding we need to reflect on our daily practices around technology and media in order to be skillful and meaningful about when, how, and *why* we use technology as part of our parenting. This knowledge helps us to avoid using screens simply as a "digital babysitter" as much as possible and instead use technology with ourselves and our children for purposeful learning, creativity, forging communities, or as a means of self-expression. Instead of using technology primarily for distraction and escape into a digital world, we can be intentional about using devices to strengthen bonds with far-away family and friends, document and relive our favorite memories and outings (the modern version of the home movie), and enhance our life in the real world. Our experience navigating technology and new motherhood is precisely what inspired us to write this book, which marries our research in technology and child development with our own experiences, successes (there are many), and learning experiences (there are also many) as a new mom.

Navigating Digital Motherhood

In some ways, the experience of motherhood has remained unchanged for thousands of years. We still strive to protect our families and do what is best for our young children. We work hard to prepare for their arrival. But in other ways, the evolution of new technology has changed how we prepare, experience, and navigate the challenges and delights of motherhood. Even before babies enter our home, the motherhood experience is impacted by digital technology. If you are a mom or mom-to-be, just think of any of the apps and websites you currently use or used while pregnant. According to recent statistics, over 50% of pregnant women use pregnancy apps on their phones or other digital devices. Apps like "The Bump Pregnancy Countdown" and "Hello Belly Pregnancy Tracker" provide information such as an expecting mom's approximate delivery date, weekly alerts on baby's size, weight, and development, and what bodily changes mom should be expecting. The app Amanda Sullivan used, for example, told her when baby was the size of an avocado, a cantaloupe, a pineapple, and a watermelon. Amanda Strawhacker used her pregnancy tracking app to help decide on a baby name with her partner. In an interview with *Women's Health*, Dr. Christine Masterson, MD, an OBGYN with Summit Medical Group in New Jersey explained this app fixation saying that, "Pregnancy is a time when your body is changing a lot, so an app can be a great way to prepare yourself for what to expect and know what's normal or not." These apps can be a helpful way to gain useful health information. But they can also be difficult to navigate in terms of accuracy, and can perhaps (for some people) add to anxiety and expectations about the different stages of pregnancy, labor, and birth. A helpful tip is to remember to use technology as a knowledge-seeking tool for yourself as carefully as you would like to use it for your child.

Social media, websites, and other aspects of digital life also impact women's mindsets, approaches, and experiences of pregnancy and motherhood. Deciding when and how to share intimate news about pregnancy journeys, IVF, pregnancy loss, and baby's arrival have been added to women's mental "to-do" list, and add to the sea of new motherhood challenges to wade through and make sense of. Seeing the social media posts of "momfluencers" and immersing in online mommy culture on TikTok, Instagram, and YouTube impact how moms view themselves, their self-esteem, their confidence, their expectations, and their own relationship with their bodies and their children.

Technology and Parenting

Once children are actually in the picture, even more questions and concerns begin to crop up for new mothers. According to research from Common Sense Media in 2018, nearly half (47%) of parents worried that their child was addicted to a mobile device. Half of parents in this research study also said they were at least somewhat concerned about how mobile devices will affect their kids' mental health. In 2025, a study by Bright Horizons showed an even higher number, with almost three-quarters (73%) of parents believing their kids needed a "digital detox" —and the majority of those parents had children under six. So if you think parenting in the digital world today is harder than parenting in generations past, you aren't alone. The Pew Research Center has also found that the majority of parents today think parenting is harder now than it was 20 years ago. Among participants in this research who stated parenting is harder today, 26% cited technology as a reason. For example, the rapid changes in technology which are hard for parents to keep up with and the influences of technology on children's behavior all make parenting challenging in ways that are unique to this generation.

Why this Book?

We're writing this book because we know firsthand the struggle that moms face when it comes to balancing technology and media with their young children—we're going through it right now with our own children. While there are a wealth of books and resources out there for parents of adolescents and older children on topics like moderating video game usage, social media, and mobile phones, there are far fewer resources out there about tools and strategies for using (and moderating) technology with very young children from birth through around age seven. In a perfect world, this might be because we may think that our children will not be exposed to *any* digital media, screens, or technology during their infancy and early childhood years. And it's true that the American Academy of Pediatrics suggests that children under 18 months have zero screen time (other than video-chatting via interfaces like Skype or FaceTime), children 18 months to age two should have limited co-viewing media exposure with a parent, and children ages two to five should have no more than one hour of screen-time per day.

However, for many families these guidelines either do not resonate or do not seem possible. This is highlighted by other Common Sense Media research showing that children ages eight and under spend an average of two and a quarter hours with screen media per day. Families with young children have a clear need, not just for more screen-time allowance guidelines that are daunting to follow, but for practical support and advice about **what** technology to use, **when** to use it, and **how** to make these technological experiences meaningful and educational for their young children.

The Questions on Every Mom's Mind

From the minute we find out we are going to have a child, most moms just want to make the right decisions for their baby. This remains true for the rest of our lives, even after they grow up. We are told that things will come to us "naturally" but worry that this may not actually be the case (and often, it isn't). Of course, modern moms struggle with the same age old questions and worries that our mothers and grandmothers before us probably had, like "how will I know if my baby is sick or just fussy?" or "what if I can't figure out how to get my baby to fall asleep?" On top of this, moms today also face a host of 21st-century specific questions and concerns related to social media, educational technology, television, screen-time, and more. This book won't help you figure out whether your baby is sick or teething (sorry!), but it *will* tackle some of the most common tech-focused questions modern moms of babies and young children face today, including:

- What gadgets or digital tools will help me keep my newborn baby safe and reduce unnecessary risks? Which ones are not worth it?
- How do I stay present when I am with my child amidst the constant pull of my phone and other digital devices?
- How do I manage my technology use around my young kids if I use technology for work?
- How do I decide where and how I share photos of my child on social media?
- What's with all the apps and online groups for moms? How (or should) I navigate this online social world for parents?
- When and how should I introduce technology to my child?

- What's with all the new talk about teaching young children to code? What can I be doing to support this as a mom? What if I have no technical background?
- What should I be doing as a mom to ensure I am raising someone who can responsibly engage with digital media?

If you are like us, your list of questions probably goes on and on beyond this selection. This book can help answer some big questions, and it will set you up for success in navigating new technology along with new motherhood. While the advice and suggestions in this book can (and should) be used by parents of any gender identity, it is specifically written with moms in mind. It taps into the aspects of pregnancy, digital mommy culture, and mom-centered social media that are unique to the experience of mothers. Additionally, it still seems to fall predominantly on mom's shoulders to manage many aspects of parenting and education discussed in this book. Research has shown that more than half (54%) of parents in households where both a mother and father work full time say that, in their family, the mother still does more when it comes to managing the children's schedules and activities. This means for many households, it still disproportionately falls on moms to think about our kids' routines, both onscreen and off. As such, this book focuses on supporting moms, but will also provide conversation starters, personal reflection quizzes, and ideas that can be used to foster discussion and collaboration with fathers, partners, and other caregivers involved with your children's upbringing. The goal of this book is to support moms and facilitate support from all those who are a part of the child-raising journey.

Seeing Technology in a New Way

This book does NOT present technology and screen-time as an enemy to avoid in order to raise happy and healthy young children. While we love and advocate for non-digital, off-screen adventures and play in early childhood, we hope this book will help you to slowly begin to think of technology in new ways. You will see that not all screen-time is created equally and we will focus on choosing tools, apps, and games that can engage young children (and parents) as *creators* rather than *consumers* of their digital experiences.

This book is set up in three parts. In Part I, we will focus just on YOU, mom! We will look at and evaluate the apps and technologies that expecting moms are bombarded with, both in tracking their pregnancies as well as in planning for your baby's arrival. Part I will also provide you with the opportunity to reflect on your own digital practices and media consumption and begin to think about a healthy media consumption plan for your family to follow once little ones are in the picture.

In Part II we will dive into the world of new motherhood and young children. We will explore the new "mommy-centric" social media apps and influences that claim to help with everything from finding playmates for your child to finding new best friends for you. We will break down the research on best practices around technology, media, and screen-time usage for young children at each developmental milestone between birth and age seven, so you will know exactly how to tackle parenting with technology at each stage of early childhood growth.

Finally, Part III will focus on how to become the tech-savvy and digitally literate mom you were meant to be. We will explore tips for travel and home use of technology, how to role-model and mentor technology usage in a way that breaks gender stereotypes, and we will think once again about how to see technology in new ways. For example, we will explore a whole host of screen-free technologies (yes, they exist!) and games that can allow you to foster fun and playful twenty-first century technical skills with your young children all without any tablets, computers, or screen time.

The ultimate goal of this book is to encourage and empower women to overcome the challenges that modern motherhood has thrown at us. As educators in the field of educational technology, a lot of our work focuses on professional development for teachers and researchers so that they can build their confidence and competence working with technology and young children. We believe in more resources for parents, and specifically moms! We hope that this book will resonate with other modern mamas, like us, just trying to find the information and support we need to do the best job we can with our little ones.

And now, as we hear our own little ones stirring from their naps, we have our cue to wrap up our own computer usage for the day (something that is a part of our current, and ever-changing, family technology and media consumption plan). Are you ready to get started on this journey navigating digital motherhood together with us? If so, try out our first self-reflection activity before moving on to the next chapter.

How Often Do You Do the Following? Check the box that applies.

Self Reflection

HOW OFTEN DO YOU TO THE FOLLOWING?

	NEVER	SOMETIMES	FREQUENTLY	ALWAYS
I worry about being too connected to my phone or other digital device throughout the day				
I check my phone during meals or other times I am socializing with my partner or friends				
I worry about how to make good use of screen-time with my young children				
I feel uninformed about deciding which technologies, apps, and media are beneficial to my young children				

If you responded "Frequently" or "Always" to most of the questions in this Self-Reflection, do not worry! This book will certainly serve to alleviate many of your worries and provide you with the knowledge and confidence you need to make great decisions around technology and media going forward. If you answered "Sometimes" or "Never" to most of these questions, you are probably already on the right track toward thoughtful and deliberate technology and media usage in your home. This book will help to keep you on the right path with specific suggestions of tools, apps, and best practices along your motherhood journey.

Wherever you are on your journey navigating digital motherhood, there is always room to grow, learn, and reflect. We're thrilled you will be coming along on this ride with us.

PART ONE

What Tech to Expect When You're Expecting

CHAPTER 1
PREGNANCY: THERE'S AN APP FOR THAT

Before even having children, technology plays a large role in the way women think about and set expectations for the experience of motherhood. For many American women in their mid-twenties, thirties, and forties, it is impossible to scroll through your social media feeds without seeing a constant flow of pregnancy reveals, birth announcements, first birthday "cake smashes," maternity photoshoots, and more. Once pregnant, technology's influence on moms-to-be only increases. Now, there is an influx of pregnancy apps for tracking your weight, symptoms, and cravings with each month that passes. There are photo editing apps for tracking your growing bump pictures and educational apps for learning about your baby's growth and development each month in the womb.

But you might be wondering, "How useful are any of these apps? How much digital tracking is helpful and when does it become harmful and take away from simply experiencing the highs and lows of pregnancy? How can expecting mothers decide how much they want to share about their pregnancies online and how much they want to keep private?" This chapter examines the ways new apps and social media have changed the way expecting moms experience their 9 months of pregnancy. It explores strategies for healthy engagement with technology and apps throughout the months before and during pregnancy.

New Technology & Preparing for Pregnancy

If you have decided to embark on a pregnancy journey–congratulations! This is one of the tenderest and most exciting stages of motherhood, and can also be one of the most fraught. There are countless digital resources to guide you in your attempts to conceive, and plenty of virtual spaces to share in the new experiences and identities that can come up along the way. In fact, most women today who are beginning their pregnancy journeys look to the Internet to find information about what foods to eat, the best vitamins to take, and to connect with other women with advice and experiences to share. In fact, the worldwide web has helped to give women today easy access to information and resources about their reproductive health like no other generation has had in the past–one of the many positive outcomes of the digital age!

Sounds like there is no downside, right? Unfortunately, with the ease of access to health information comes the ease of access to misinformation, targeted ads, and opinions that are not grounded in medical science. Nina Jankowicz, a disinformation researcher, wrote all about this in her opinion piece for *WIRED* in 2022 called, "The Internet is Failing Moms to Be." Jankowicz writes that popular pregnancy apps "capitalize on the excitement and anxiety of moms-to-be, peddling unrealistic expectations and even outright disinformation (malicious and deliberately inaccurate information) to sell ads and keep users engaged," and that these apps "foster negative repercussions on the physical and mental health of both mothers and their unborn children."

And it does not stop at pregnancy apps. Research published in the journal *Contraception* found that most crisis pregnancy centers listed in U.S. state resource directories for pregnant women provide misleading or false information regarding the risks of abortion. Misinformation is also shared through well-meaning advice and opinions on social media or other online forums. Many of us can recall all too vividly the misinformation related to the COVID-19 pandemic, vaccines, breastfeeding, and pregnancy just a few years back.

Despite all of the risks to misinformation and other potentially negative impacts, women *can* safely and wisely navigate websites, social media, and apps effectively during their conception and pregnancy journeys. Let's jump into specific examples and resources for each stage of your pregnancy and conception journey.

Pregnancy in the Digital Age

Mom's mental, physical, and social wellbeing are paramount for a healthy pregnancy journey, and it can really become a lot to keep track of! This is where new technology, mobile apps in particular, have become a very handy tool. In fact, some researchers estimate that more than half of all pregnant women download a pregnancy app. So it may not be surprising that some researchers have found that pregnancy apps are the most used health app–even more popular than fitness apps.

Both of us writing this book (Amanda Sullivan and Amanda Strawhacker) had fun using different pregnancy apps during our own pregnancies. Using the free Ovia app, for example, we were able to get regular updates about our babies' growth and our own changing bodies, and forums where we could chat with other expectant mothers who had similar projected due dates. These pregnancy apps also offered text and video articles to address typical questions like "when will my morning sickness end?" or "what are the signs I am in labor?" that many expectant mothers are wondering about. While the health information was often useful, the pieces we liked most were the fun and cute elements of these apps. For example, knowing what week our baby is the size of an avocado or a watermelon with fun graphics that can be shared with friends or on social media.

For many others, the health information provided by these pregnancy apps are what makes them useful. This was certainly magnified for many pregnant women during the COVID-19 pandemic, when appointments were being canceled and rescheduled left and right. Even before the pandemic, researchers Sayakhot & Carolan-Ola found that pregnant women often sought information on the Internet and did not discuss much of what they found with their physicians. This becomes a potential problem when it is not always clear whether the information provided in popular pregnancy apps is entirely medically accurate. A recent research study published in the *Journal of Medical Internet Research* surveyed pregnancy apps and found over 60% did not have comprehensive information for every stage of pregnancy and only 28% cited medical literature.

So what does a pregnant woman in the digital age do to navigate the apps and virtual resources out there? Should you abstain completely from using a pregnancy app? We do not believe the answer is black and white. Often, the internet is the first place moms look when in need of some clarifying information. For example, one mom we know told us that after her gestational diabetes diagnosis, she felt she knew nothing about how to care for herself and found digital resources invaluable to teach her how to count carbs and plan recipes that would help her eat in a way that her body needed. But, it may be helpful to remember that often these apps are created for a variety of reasons, including but not limited to making a profit. You may learn wonderful new facts and tips - but you should always check with a medical healthcare provider before taking action on any information posted on pregnancy apps or websites. For more tips and tricks for navigating the world of pregnancy apps, check out our Do's and Don'ts in the table below.

THE DO'S AND DON'TS OF USING
Pregnancy Apps

DO...	DON'T...
CHECK ANY ADVICE OR MEDICAL TIPS FOUND IN PREGNANCY APPS WITH YOUR OBGYN OR MIDWIFE	RELY ON INFORMATION PROVIDED IN THESE APPS AS YOUR ONLY SOURCE OF MEDICAL INFORMATION
HAVE FUN WITH GRAPHICS AND IMAGES ABOUT BABY'S SIZE AND WEIGHT	TAKE INFORMATION ABOUT THE PROJECTED SIZE OF YOUR UNBORN BABY TOO LITERALLY
HAVE FUN READING TIPS AND ADVICE FROM OTHER MOMS ON COMMUNITY MESSAGE BOARDS	ACT ON ANY ADVICE OR TIPS POSTED IN COMMUNITY MESSAGE BOARDS WITHOUT CHECKING WITH A MEDICAL PROVIDER
LEARN ABOUT YOUR CHANGING BODY THROUGH VIDEOS AND ARTICLES WITHIN PREGNANCY APPS	USE ANY INFORMATION FOUND ONLINE AS A SUBSTITUTE FOR REGULAR PRENATAL HEALTHCARE

Infertility, Pregnancy Loss, and the Internet

The Struggle of Trying to Conceive & Social Media

We asked moms how social media and the Internet influenced their journey into pregnancy and motherhood. Here is what some of them had to say:

"I did 5 years of IVF and it was hard to see people grow their family while I struggled. I unfriended lots of acquaintances. I also got support from IVF groups."

"While I was fortunate to only ttc [try to conceive] for 3-4 months, it was upsetting at times to see other pregnancy announcements during that time."

"As someone who had an infertility journey and was unable to have a second, social media announcements of pregnancies and births were - and still are - incredibly difficult. They often come as a surprise - you're just scrolling along and bam!"

Thus far, we have been exploring the ways that new technologies, particularly mobile apps, impact women during their pregnancies. But for many women working to start or grow their families, pregnancy does not always go as planned–or even happen at all. *Infertility* is defined as the inability to conceive children and it impacts more women than you may think. According to the Centers for Disease Control and Prevention (CDC), in the United States, about one in five heterosexual women aged 15 to 49 years with no prior births are unable to get pregnant after one year of trying. The CDC also states that about one in four women in this group (26%) have difficulty getting pregnant or carrying a pregnancy to term- this is defined as *impaired fecundity*.

Digital resources offer a means of helping support women along their pregnancy and infertility journeys. Recent research published in *BMC Women's Health*, for example, found that participating in infertility forums online (such as on sites like Reddit) provided study participants information about fertility treatments and social support in the process of coping with infertility. The forums offered participants information, solidarity, and the opportunity to give and receive support from individuals who could understand what they were going through. However, it is worth noting a potential negative aspect that researchers found, that individuals participating in online infertility forums risk taking in information that is not evidence-based or rooted in medicine. This may sound familiar, since we just discussed misinformation and advice not rooted in medical facts in the previous section of this chapter on pregnancy apps. We hate to break this to you, but get ready to hear that a lot throughout this book. For every positive aspect of websites, social media, and apps, there is also the downside of inaccurate information being shared. Learning to parse through the information provided is a key aspect of being a tech-savvy mama in the digital age.

How Many of These Fertility Forum Acronyms Can You Recognize?

There are a growing number of online message boards, social media pages, websites, and other online communities populated by those TTC (Trying-to-Conceive). These online communities can be a wonderful source of support and encouragement if you're coping with infertility or simply beginning your journey trying to get pregnant. But the acronyms and jargon found in these spaces can be overwhelming and feel like a new language when you are first venturing into these online spaces. Don't be intimidated! We'll break down a few of the most common ones you'll come across here, and you can find longer lists on most popular pregnancy and family planning websites.

AF - Aunt Flow (menstrual cycle)
BFN - Big Fat Negative (negative pregnancy test)
BFP - Big Fat Positive (positive pregnancy test)
CM - Cervical Mucus
DE - Donor Eggs
DPO - Days Past Ovulation
TTC - Trying to Conceive

Digital resources and social media also offer a new experience of community and support for women experiencing miscarriage. According to the American Pregnancy Association, miscarriage is quite common, occurring in 10% to 25% of known pregnancies. In other words, if you know at least ten women of childbearing age, you probably know someone who has had a miscarriage—whether or not you are aware of it. Having a miscarriage is a hugely heartbreaking and often traumatic event. Still, most women do not receive any mental-health support as part of the medical management of their miscarriages. And while things have been changing in recent years, talking about miscarriage still feels "taboo" for many women. This means many women who experience miscarriages end up enduring their mental and emotional suffering alone.

Silence around the topic of pregnancy loss has also resulted in misconceptions about miscarriage, infertility, and pregnancy loss. For example, a research study conducted by Bardos, Hercz, Friedenthal, Missmer, Williams in 2015 surveyed over 1,000 American adults. The researchers found that more than half of participants (incorrectly) believed that miscarriage is uncommon, and nearly a quarter said they believe that lifestyle choices are the most common cause of miscarriage. Of those who had a miscarriage, 37% felt they had lost a child, 47% felt guilty, 41% reported feeling that they had done something wrong, 41% felt alone, and 28% felt ashamed.

Here is where the Internet, and social media in particular, has been making a notable change in this experience for women in recent years. Research published in the journal *Paediatric and Perinatal Epidemiology* in 2020 analyzed trends behind the impact of celebrities' public disclosures of miscarriage on Twitter and uncovered interesting patterns. The study, titled "Discussions of miscarriage and preterm births on Twitter," found that postings about pregnancy loss increased when celebrities like Michelle Obama or Gabrielle Union went public with their own miscarriages. As more celebrities have used social media to share their personal experience with a once-taboo topic, many more women have found the voice to share their own stories, sometimes months or years after the experience of miscarriage itself.

On other social networking websites like Facebook, community groups offer women a place to share their stories, offer advice, and ask questions. Popular Facebook groups such as Miscarriage Mamas and Miscarriage and Pregnancy Loss, both of which currently have over 147,000 followers at the time of writing this book, offer digital communities where women seek medical advice and suggestions on how to navigate their experiences with pregnancy loss. These groups offer women a sense of community and support during what can be an extremely isolating experience.

Dissecting the "12-Week Rule"

We could not write a whole chapter on pregnancy, miscarriage, and the digital age without discussing the so-called "12-week rule" for pregnant women. This "rule" refers to a long accepted convention that you should not publicly announce your pregnancy until three months in, because after that time the chance of miscarriage drops significantly. According to trends from the National Survey of Family Growth over the last two decades, the average woman learned about her pregnancy in week five, meaning most women would need to keep their news a secret for nearly two months if following the 12-week rule. While some women strictly abide by this rule, many have a hard time waiting an entire trimester to discuss this huge life transition with friends and family, and others disagree with the concept on principle. Many women believe this rule reinforces the damaging idea that women should only share "happy" news about their reproductive journeys and that any experiences of pregnancy loss should be suffered alone.

If you are pregnant and considering when to share your news, take some time to reflect and answer the following questions with your partner (or anyone else involved with supporting you during your pregnancy):

- How would you feel experiencing an early pregnancy loss if you had not yet told any loved ones you are pregnant? Would you be able to get the support you need?
- How would you feel announcing your pregnancy to friends and family and later losing the pregnancy or facing medical challenges? Would you feel supported or unnecessarily pressured to discuss your situation?
- Is there a happy medium that might be right for you? For example, telling a few trusted friends and family members prior to 12 weeks?

Whatever you decide, remember that this is **your** pregnancy, and no societal rule or convention should dictate when or how you choose to announce your news. Think about what feels right to you, your partner, and any involved family members. Want to share your pregnancy on social media the minute you get that big fat positive test? GO FOR IT! Want to keep your pregnancy under wraps for as long as possible? GO FOR IT! There is no one "right" way to go about sharing such personal and intimate news–on—or offline.

Finding Comfort Online

"After I had a miscarriage, I felt sadness, loss, and grief like I had never known before. This sadness was magnified by feeling utterly alone: I had been trying to follow the '12-week rule.' Because of this, I had not told any family members I was pregnant and had only told two trusted friends, both of whom lived far away. I can still remember lying in bed in the hospital wishing I could call someone to come be with me... but knowing that calling them meant I would not only have to explain my miscarriage, but also that I had been pregnant to begin with.

As the days went on, I had to work and care for my son as if nothing had happened to me - because no one knew what happened to me. In my loneliness, I found comfort online. I began to read about women's experiences with miscarriage and learned I was not alone. I read statistics and learned how common my experience was, even though we are often pushed not to talk about it. I found comfort in support groups on social networking websites like Facebook. When I was pregnant once again with my daughter about six months later, I decided to throw the 12-week rule out the window.

I made the choice to savor and find joy in every possible fleeting moment of my pregnancy, sharing my news just weeks after a positive test. This might not be the right decision for everyone, but it was for me and my family."

- Amanda Sullivan

Birth and Postpartum Support

If you picked up this book, we are guessing that you have, or are close to having, a baby or young child. After nine-(ish) months of waiting and growing, you may be feeling joy at the prospect of meeting your baby, anticipation to finally reach the end of pregnancy, or eagerness for what will come after. One emotion almost all moms face is the mix of excitement and anxiety that comes with actual childbirth. For many mothers, this next step of the parental journey is as beautiful as it is overwhelming. Most of the subsequent chapters in this book will be specifically focused on your baby and young children. But for now, we want to focus on YOU, the mother, and your own well being–something that we as mothers often forget to think about and devote attention to.

The March of Dimes, a long-established non-profit organization focused on maternal and infant health, estimates that up to 80% of women experience the "baby blues" after giving birth. Despite the innocuous name, the "baby blues" can include intense feelings of sadness, worry, and exhaustion that often emerge in the first few days after having a baby. For most people, the baby blues resolve naturally in a few weeks without any additional treatment. For others, this may be just the beginning of more serious postpartum depression, which can last up to a year or longer. According to the CDC, about 1 in 8 women with a recent live birth will experience symptoms of postpartum depression. Regardless of whether you experience the baby blues or postpartum depression, the transition into new motherhood (or to welcoming an additional child) takes a toll on a mother's mental and emotional wellbeing.

In some ways, mothers today are lucky. The Internet offers easy access to mental health resources, virtual appointments with therapists and social workers, and facts and information about mental health at your fingertips - something no mothers of past generations had access to. On the other hand, social media can also fuel feelings of anxiety, depression, and feelings of inadequacy during the initial months (and longer) after giving birth. For example, researchers have found that mothers who frequently compared themselves to others on social media sites felt more depressed, overwhelmed, and less competent as parents. In other words, if you are a new mom who is struggling with seeing (and inevitably *comparing* yourself to) picture-perfect moms living apparently effortless and beautiful lives on Instagram, this engagement may make you feel worse about yourself. It may be helpful to remember how easy it can be for social media to create a veneer of perfection while masking some of the more mundane realities of diaper blowouts, sleep deprivation, and stress that *all* moms (yes, even influencers) face. Still, this is when turning off your phone and computer may be the best choice, so you can take a "digital detox" while you work on your own mental health and bond with your baby.

Did You Know?

You have probably heard of the three trimesters of pregnancy. But have you ever heard the term, "the fourth trimester" before? The fourth trimester is a period coined by pediatrician Dr. Harvey Karp and refers to the first 3 months of a baby's life after birth. This is a time when babies have regular routine checkups and appointments but mothers typically have just one postpartum check - usually 6-8 weeks after giving birth.

The fourth trimester is a very vulnerable time for mothers and their families, full of extreme physiological, social, and emotional changes. During the fourth trimester, social media and the Internet can be an additional strain, particularly for new moms. Mothers in the fourth trimester should use social media for fun, connection with friends and family, and to share baby and personal updates that they feel comfortable sharing. When social media use contributes to feelings of inadequacy, body image issues, comparisons to other moms who seem to "have it together," over-distraction from baby, or loneliness, moms should consider scaling back their usage.

Important: *If you or someone you know is in a serious crisis, tell someone who can help immediately. Call 911 or go to the nearest hospital emergency department for emergency medical treatment. You can also contact the confidential and free National Maternal Mental Health Hotline, 24/7: Call or text 1-833-9-HELP4MOMS (1-833-943-5746).*

Digital Tools and Resources

Now that you know about the diversity of virtual communities, apps, message boards, and online resources out there, how can you narrow down the handful that will support you on your pregnancy and postpartum journey? You may find that you can access excellent virtual communities on favorite social media platforms that you may already be comfortable using, like Facebook, Instagram, and TikTok. On the other hand, knowing there are so many digital resources out there may also feel overwhelming. Access to so much information, statistics, and advice can make you feel anxious or stressed. Remember, digital tools are there to support and educate you - not to add to your stress. Try to avoid information overload and googling every pregnancy symptom. Instead, focus on seeking specific resources or information that can help you feel healthy and confident.

Access to Info in the Digital Age

"In some ways, access to digital resources during my pregnancy was amazing. I felt empowered, educated, and prepared with questions and hints about what to expect at every stage and trimester. In other ways, it was kind of scary...COVID restrictions on in-person birth classes and hospital tours forced me to feel like I was learning to give birth by watching YouTube videos. My partner wasn't allowed to attend most of my OB appointments, and sometimes I misunderstood or misinterpreted labwork info I received through my digital OB app.

After that experience, I really believe that the best thing to do is keep your real-life support network of doctors, friends, and family in close contact (video and phone calls are enough!) and don't fall into the trap of trying to absorb all the information alone or at once. The reason all these digital experiences exist is to help us connect with others going through similar experiences, and grow our network of supportive individuals - not to enforce impossible standards on ourselves and our pregnancy journeys."

- Amanda Strawhacker

If you are looking for more resources to support you during and directly after your pregnancy, check out Appendix A, where we provide a roundup of our favorite types of digital resources to support you at this magical time in your motherhood journey. Keep in mind that specific apps and websites change in relevancy from year to year. Some of our favorite apps may (sadly!) not even exist by the time you happen upon this book. Instead of using this list to find specific apps or websites, we suggest you use this list to spark ideas for you on what types of resources are out there and where you might find them. Remember when exploring resources to think like a detective: check for *who produces the information* (well-established government, non-profit, and research organizations are usually best informed, followed by accredited professional individuals), and *how they arrived at their recommendations* (standardized research trials? Personal experience in a hospital or midwife setting?). Whatever information you find, always keep in mind that your own situation is unique to you, and feel free to voice any questions or concerns with your support network of loved ones and medical professionals.

Deciding What to Share Online

You are now familiar with pregnancy apps, online communities, fertility forums, and the range of community groups on social networking websites like Facebook. We have seen that many people gain support and a sense of community by discussing aspects of their pregnancy journeys online, explored tips to evaluate prenatal information sources you may find and share on the Internet, and discussed some factors to consider when deciding when to announce your pregnancy news. But what about after your baby is born? How do you decide what to share about your baby and family online? This is a very big question, and as such, we will revisit discussing this question in different places throughout this book. Beginning in the next chapter, we will dig into your own personal digital practices and how you engage with social media and decide what to share about yourself. These will be our building blocks and foundation for thinking about how an infant or child complicates what you decide to share online (and how you share it).

CHAPTER 2
REFLECTING ON YOUR OWN DIGITAL PRACTICES

Many moms (both of us included) begin to really think about managing technology and screen time for the first time when their children enter the phase of being interested in tablets, phones, and televisions. This usually happens when children are somewhere in the age range two to five years. Often overlooked are *parental habits* when it comes to screen time management and usage. The uncomfortable truth is, while parents are busy worrying about their children's technology and media habits, they are often sending mixed signals to their children anytime they check a text during family dinner, answer a personal phone call in the car, or turn "background television" on while cooking or cleaning the house.

Whether we like to admit it or not, the way we as parents interact with our smartphones and other technology directly impacts our relationships with our children. According to research conducted by the Pew Research Center in 2020, 68% of parents in the United States say they at least sometimes feel distracted by their phone when spending time with their kids. The good news is it is easy to make a few healthy changes in your personal technology and media consumption. While the ideal time to make these lifestyle changes are *before* your baby arrives (add it to your baby preparation checklist!), moms at any point in their parenthood journeys can benefit from reflecting on personal technology usage, and it is never too late to make some positive changes.

In this chapter, we explore some of the ways adults engage with technology, and how these practices might (or might not) change with the arrival of a new baby and an ever-evolving family dynamic. We will take time for self-reflection on our own digital practices and begin working towards a balanced and sustainable technology and media plan to help you feel prepared to model healthy tech habits for your children in the years to come.

The Newborn Parent Survival Phase

If you are getting ready to welcome a new baby into your household, you may already be bracing yourself for the initial phase of sleep deprivation, exhaustion, and major personal adjustment to new motherhood. During this time, many new parents use television or scrolling on their phones to help keep themselves awake and alert when caring for newborns who are dealing with colic, illness, or simply do not want to stay put in their cribs through the night (both of us writing this book have been there!). Additionally, many mothers who find themselves nursing or bottle feeding when the rest of the house is fast asleep use television and other technology to help combat the loneliness of those quiet late-night hours.

If you are worried about how your interaction with technology during feeding or nap time will impact your baby, we're here to reassure you that (almost) anything goes during this initial "survival phase" of parenthood - so do whatever you need to do to keep your sanity! Research tells us that newborn babies (zero to three months) are not yet able to filter information coming from all of their senses the way adults can. This means they really are not able to focus on what's on the screen itself. So when you welcome your baby, if you find yourself sending a few texts or emails or watching TikTok videos on your phone, you can rest assured that your newborn infant is unlikely to be picking up any direct information from these devices.

All that said, it is important to note that adult behavior changes when technology is added into the mix. Technology causes adults to divide their attention from engaging with their babies and children, offering less time for play and learning. In fact, a lot of research has shown that when television is on, even in the background, the number of words a parent speaks to their child decreases. Over time, this means children are getting less active conversation with parents, and might eventually delay language development.

Exposure to background noise may also distract children from their own playing and learning. We will dive into the impacts of screen-time on young children further in Chapter 5, but it is worth mentioning now as we are thinking about why it is important for expecting parents to reflect on their digital habits *before* adding infants or children into the mix.

Did You Know?

Research has spoken! Background media may seem harmless, but it can have potentially negative effects on babies, young children, and families. Here are three research findings every parent or parent-to-be should be aware of:

In the presence of background TV, the number of words and utterances spoken per minute by the parent decreases. (Pempek, Kirkorian, & Anderson, 2014).

Currently, there is not enough research to distinguish any difference in spoken words per minute in the presence of other kinds of background audio (e.g., podcasts, music). If you want to minimize any chance of baby confusing your spoken language with words from the audio background, you could try sticking to music without lyrics.

Background TV disrupts infants' and toddlers' toy play. (Setliff and Courage, 2011; Schmidt et al., 2011)

Background TV decreases high-quality parent–child interactions. (Kirkorian et al., 2009)

Technology Self-Check

"In today's digitally connected world, parents have to manage their own relationship with the internet and mobile devices, along with managing their children's use of and exposure to the same technology. This brings with it a host of both benefits and challenges."
-2020 Pew Research Center report on parent attitudes on digital technologies

Many of us never stop to think about our personal technology and media habits before having kids because we have no need to. After all, why would you care how much television or texting you do as an adult unless it was negatively impacting your work or personal life? What we might forget is that tech habits that are healthy and fine in a household with just adults are not necessarily healthy and fine in a household with babies and children. It is MUCH easier to break habits and make shifts in your schedule *before* you add in the major upheaval of becoming a first time mom. This is why we highly recommend you check in on your technology consumption as part of your prep for welcoming a baby. Nevertheless, this is still helpful to do at any stage in your motherhood journey.

So where do you begin? One way to start is to think about a typical day in your personal life and complete our Technology Consumption Self-Check. Try not to have any judgment about your habits and answer as honestly as possible - no one is reading this except you! On this self-check you will reflect on how much you text, make phone calls, play digital games, go on social media, watch television, and more. We will also ask you to think about household habits like having television on in the background while you do other activities.

Remember, there are no "right" or "wrong" answers here. The goal of this exercise is simply for you to get a sense of your current habits and digital activities. Are you ready to check-in on your habits? Complete the Self-Check now!

Technology Consumption Self-Check

How often do you use your phone for texting each day?

Never — Once or twice — Three to five times — Six to ten times — Too often to count

How often do you use your phone for calls each day?

Never — Once or twice — Three to five times — Six to ten times — Too often to count

How often do you use your phone for games or apps?

Never — Once or twice — Three to five times — Six to ten times — Too often to count

How often do you use your phone to engage with social media (e.g., TikTok, Instagram, Facebook, etc.)?

Never — Once or twice — Three to five times — Six to ten times — Too often to count

How did you answer the questions on the self-check? If you selected that you engaged with at least half of the digital activities "Too often to count" or on the higher end of the range in a typical day, you are just like many other adults out there! Recent survey data collected by the Pew Research Center shows that about seven out of ten Americans say they use some kind of social media site, a share that has remained stable over the past five years. The same research also demonstrated that the majority of Americans say they use YouTube and Facebook, and that use of Instagram, Snapchat and TikTok is especially common among adults under 30. Maybe you stated that you use social media like Instagram multiple times a day. Once again, this couldn't be more normal! Surveys collected by the Pew Research Center also showed that the majority of adult Instagram users in the U.S. say they use the site at least once a day, and 38% say they use the site several times per day.

You may be feeling very happy about how you responded to this self-check. On the other hand, you might find yourself surprised by some of your answers. However you answered the questions or are feeling at this moment, we want to congratulate you on taking the first step to finding balance with digital technology and media usage in your personal life. As you are looking at your responses to this self-check, think about the following:

- Did any of your answers surprise you?
- Are there any digital practices where you would like to reduce your daily usage?
- Will any of your digital practices impact how you connect and spend time with your baby/children (or do they already)?

We suggest that you encourage your partner, co-parent, grandparents, or any other family members and caregivers directly involved with your child's care to take this self-reflection so that you can discuss your answers and habits together!

Did You Know?

Research from the Pew Research Center has found that:

- More than half of parents who have at least one child under the age of 18 say they spend too much time on their smartphone (56%).

- Mothers (61%) are more likely than fathers (49%) to say they spend too much time on their smartphone.

- A majority of parents (68%) who have at least one child under the age of 18, say that they at least sometimes feel distracted by their smartphone while they are spending time with their children, including 17% who say they feel this way often.

Creating a Healthy Technology and Media Plan

In Part II of this book, we will talk more about how you can support a healthy media and technology plan for your children, but right now we are staying focused on you and any other adult caregivers involved with your baby or children. Just like with exercise and food, children will learn healthy habits much more from watching what you do than listening to what you tell them to do. Habits of any kind can be hard to change, which is why we recommend making a few healthy shifts before welcoming a baby, if you can (but it is never too late to start).

Having a plan will help you stick to the healthy goals you want to keep. Here are some things to keep in mind:

- **Think about Your Priorities** - What are your priorities when it comes to managing your own digital media and technology time? Is it reducing your overall time spent on screens? Having more uninterrupted time with your partner? Everyone's goals and priorities will be different. Having clear goals and priorities in mind will help you develop a plan that works for you.

- **Healthy Practices** - Think about how your technology and media usage can support healthy balance in your life. If scrolling on your phone is impacting your sleep, for example, this might be an area that needs shifting. Any media usage pattern that impacts your healthy eating, sleeping, exercising, and mental health is worth reflecting on and perhaps changing.

- **Screen-Free Zones** - Some media plans may focus on overall time spent using digital technology. But for others, simply designating "Screen-Free Zones" in the house is enough. For example, would you like your bedroom to be screen-free? Baby's nursery? The dining room? Keep in mind that for Screen-Free Zones to be effective, all adult caregivers in the home need to follow the same guidelines.

- **Screen-Free Times** - For others, "Screen-Free Times" work better than zones, especially for those living in smaller spaces where each room may be used in multiple ways throughout the day. For example, family dinner or breakfast might be a time where screens are never used. Or, bedtime might be a time of day that you always commit to making screen-free. To help keep you honest, there are apps that can track your device usage, block any apps you choose or limit internet access during set time frames, or remind you to turn off your device after a certain amount of time has passed.

Setting your priorities and agreeing to a plan for screen-free times of day or zones in the house before your baby's arrival will set you up for success once the baby arrives. Breaking pre-motherhood habits of all kinds can be hard! One mom told us that she even had to make an adjustment to her normal daily habit of watching the news every morning to catch the weather, since her three-year-old daughter now picks up on some of the headlines that aren't child appropriate. Be forgiving and patient with yourself as you adjust to a new normal. Remember, the plans you will stick to long term should have flexibility and match your lifestyle, rather than feel stifling or unrealistic. There may be times when you or your kids are sick, exhausted, or simply need to deviate from your plan for whatever reason - and that is 100% okay! Just as extreme diets rarely work for those trying to eat healthier, the same is true for extreme shifts in your technology consumption. Work to make a few healthy changes, but remember that one indulgent day (or week!) will not make or break your overall healthy media habits.

Breaking Pre-Motherhood Digital Habits

"Looking around our family room, I can already count four devices within easy access: TV, video game console, smart phone, and laptop. Any of these devices might compete for my attention-attention I could also pay fully to the baby in my arms, for example, with eye contact during nursing, or reading a book together before naptime. Making the choice to spend this quality time with my baby might sound obvious. But if I'm being honest, it takes practice to resist my pre-motherhood habits of constant tech engagement! The upshot is that my baby is far more entertaining than any show or app I've ever found, so that always helps me remember why I actively remind myself to put down the phone."

- Amanda Strawhacker

"Prior to having children, I was someone who loved to have 'background' television on at all times. While I cooked, cleaned up, or got ready for work, I loved to have a movie or TV show playing, even if I was just listening to it. This was a big change for me after having kids! But my husband and I made a meaningful choice never to have background media on once we had kids. This was a really positive decision for us, and something we have been able to stick to for five years and counting. Now, I look forward to alone time when the kids are sleeping to catch up on news or TV shows. I've found that I actually appreciate my screen-time more now than when I consumed it in the background of other activities."

- Amanda Sullivan

Only you know what will work best for you and your family, but most children benefit from little to no screen-based technology use in the household under age two, including "background" television that is on for parents. You can start getting used to habits that can limit (or eliminate) passive screen-time and mindless tech use in your home, allowing you to more fully focus on your baby when they arrive. Luckily for new parents, babies spend most of the day napping so even if you start planning your healthy media consumption plan *after* baby arrives, you can get used to your new habits slowly over time, integrating more and more active playtime and technology breaks during baby's waking hours.

Next Steps

Okay, we know you probably weren't expecting homework when you picked up this book, but SURPRISE! We have a few small tasks to get you set on the path to success:

- **Record & Reflect** - Using the pages provided at the end of this chapter, record your personal technology/media usage during one 24-hour period. Note the time, duration, and type of technology/media you used. What did you notice about the type of technology you used and when you used it? Are there any places you think you might want to cut down on usage? If you are expecting a child, are there any specific places in your technology routine you will need to adapt or eliminate before a baby is in the picture?

- **Talk with Your Partner / Co-Parent** - Encourage your partner or any other adults who will play a large part to engage in reflection and conversation with you about their healthy media and technology consumption. To get them started, use the blank copy of the Technology Consumption Self-Check at the back of this book. After discussing and comparing your answers on that self-check, encourage them to do their own 24-hour recording and reflection too. Are there any things you both agree needs shifting? What about any areas where you are already feeling confident about your balance and healthy habits? Remember to celebrate strengths too, rather than simply focusing on areas of improvement!

- **Draft Your Healthy Family Media Usage Plan** - Work with your partner, co-parent, or other adults who will be heavily involved with childcare on drafting a family media usage plan with you that works for your household—and start to follow it now! But remember, these plans should be considered flexible and subject to change as your family grows and changes over the years. When your kids are old enough, you may wish to engage them in regular planning sessions where you revise and adapt your family plan.

Need help getting started? The website HealthyChildren.org has a great interactive resource for drafting a healthy media consumption plan! Just search "Family Media Plan" on the homepage to find lots of useful tools, articles, and resources!

What's Next?

Now that you have taken some time to reflect on your own pre-parenthood (or pre-new-baby) technology and media habits, with a focus on areas where you may wish to cut down, you are ready to start thinking about some new technologies you probably didn't encounter before having kids. In the next chapter we will explore the newest apps, tools, and technologies that may actually *help* you and your baby during your initial year of parenthood. Read on to find out which digital tools you should be adding to your baby preparation checklist, and which you may be better off without. As you learn about new tools, feel free to return to your technology and media consumption plan to revise and make changes!

Choose a day when you are not working outside the home. Any day when you are full-time with your whole family and/or just your children at home works great. Keep track of your technology habits from morning to night for a full day!

Note down when you look at your phone, log into your computer, etc.

What trends do you notice? Are there any areas you wish to cut down?

Log of Technology Use for One Typical Day with the Family

Date:

Waking Up (6am-9am)

Morning (9am-12pm)

Mid-Afternoon (12pm-3pm)

Late Afternoon (3pm-6pm)

Evening (6pm-9pm)

Night (9pm-Bedtime)

CHAPTER 3
NEW TECHNOLOGY & PLANNING FOR BABY'S ARRIVAL

Pregnancy and babyhood are some of the most precious times in your parenting journey, but can also feel the most overwhelming! Of course you will probably find yourself dreaming about matching "mommy and me" outfits and adorable baby clothes (and hats, and bow ties, and socks…), but when you get to planning purchases for the nursery, it's helpful to think practically. How can you plan ahead to adapt your home and lifestyle to welcome a new baby into your family? What items do you really *need* for your new little housemate, and which ones can you skip? And most importantly, how can technology help?

You will probably find that your baby spends most of their time on these "big four" activities: feeding, sleeping, elimination (diapers), and enrichment (play). Exploring and regulating these behaviors is the developmental "work" of a young infant, and as any parent can attest, it takes a lot of effort and practice from the whole family to help their baby along! With any newborn, parents face hurdles that may be completely developmentally appropriate (ready for those 3:00am night wakings?) and others that might require outside guidance to support baby and mom (for example, there's a reason breastfeeding is often called a "journey"). It's important to remember that no one can control how a baby experiences their big four activities. However, you *can* use strategies to track behavior, predict milestones, and plan logistics to make their developmental transitions and your parenting responses as aligned as possible. That's where technology comes in!

Getting Ready for Baby Self-Check

Using data-tracking apps to track baby's feeds.

Very Comfortable — Comfortable — Neutral — Nervous — Very Nervous

Collecting and storing breastmilk or formula.

Very Comfortable — Comfortable — Neutral — Nervous — Very Nervous

Using video/sound monitors while the baby sleeps in another room.

Very Comfortable — Comfortable — Neutral — Nervous — Very Nervous

Tracking sleep patterns and "wake windows" to predict baby's sleep habits.

Very Comfortable — Comfortable — Neutral — Nervous — Very Nervous

How did you answer the questions on this self-test? If you are welcoming your second, third, or fourth child, you might have selected "comfortable" or "very comfortable" with most of the questions in this self-check. But even experienced moms know every baby has different needs, and new recommendations and research are coming out all the time. There are dozens of new tools, apps, and online resources designed to support mom with all the changes that come with a new addition to the family. In this chapter, we've organized many of these resources according to the common baby behaviors of feeding, sleeping, elimination, and enrichment. Many of these tools can include other people in your baby's digital network of care, and can support all areas of baby's development over the first zero-to-two years of life! You can decide what works for your family's needs.

Feeding

Today, there are tech tools to support breastfeeding, bottle-feeding, pumping, and more. It's a wide world of baby feeding out there, but most options basically fall into two categories: breastfeeding and bottle-feeding. Either of these approaches can be a great way for your family to prioritize nutrition and bonding with your newborn. You may already know what approach you want to pursue for feeding your baby, or you may be open to exploring many options. You can learn more about feeding at your obstetrician, hospital, or online–see Appendix A for some web resources we love.

For many moms, breastfeeding will be their baby's primary feeding strategy. Breastfeeding has so many benefits, including protective health and immunity impacts for mother and baby, but one of the most convenient benefits is that it's free and requires no special technology to start! However, you may find that you want to purchase *some* technology for your own convenience, or during necessary separations from your baby. New products for breast health are also coming out, such as breast warmers and massagers to increase milk production and soothe clogged ducts. Breast pumps are tools used to express breast milk into bags or bottles, and the technology has come a long way since it was first invented in the 1850s! Moms can now explore hands-free and cordless pumps that let you multitask while pumping, and some even fit inside your bra and run silently so you can pump while doing chores or having meetings. If you know you want to try breastfeeding, it's a good idea to at least acquire a breast pump before the baby's arrival. Fortunately, since the American Academy of Pediatrics recommends exclusive breastfeeding for the first six months of baby's life, US insurance companies must cover the cost (purchase or rental) of a manual or electric breast pump. Not all pumps are covered (including, for example, hospital-grade pumps, which often require a "medical necessity") so you might decide to invest in a specific pump based on your own preference or convenience. When comparing pumps, don't forget to look into compatible bottles or bags for storage, and any accessories you might need to customize your pump. Another idea from savvy moms we know is to explore and collect formula samples before the baby comes home. Even if you plan to breastfeed, no one can predict when their baby will require a special diet or a higher milk supply than what mom can produce, so it can be a relief to feel prepared in those early days.

Do I Need a Lactation Consultant?

"My daughter took to breastfeeding very quickly, but when I went back to work I was so confused about pumping! I knew nothing about all the parts - what the heck are flanges and duckbills? I was worried I was going to feel like a cow at a dairy factory and was so sad not to just be able to nurse for every feed, something my daughter and I both enjoyed and found very comforting and sweet. Whenever I practiced pumping at home, I was also disheartened by how little milk I was getting.

Eventually I contacted a lactation consultant that a friend recommended. **I didn't know that Lactation Consultants (LCs) can help with <u>anything</u> related to baby feeding, even beyond breastfeeding. Not only was my in-house visit covered by insurance, but I gained access to a health portal app to communicate with her for free, for a whole year after my first consultation.** I learned so much about pumping efficiently, using correctly sized pump parts, and storing milk safely that within a week my "stash" of frozen milk doubled, and now I have no problem pumping at work while my daughter is at daycare. Even better, the time we spend breastfeeding at home during nights and weekends is still special and intimate, and I am still able to reach out to my LC anytime a new question pops up."

-Amanda Strawhacker

When it comes to technology for preparing frozen milk or powdered formula for babies, moms have more options than ever before. There are high-end formula dispensers that work like a coffee machine to mix the number of ounces you need at specific proportions and temperatures, or you can choose lower-budget hand-mixing with non-technological measurement and storage products. There are also machines designed to heat breast milk to optimal temperatures to preserve nutrients in the milk, and machines that automatically shake bottles of formula so no powder clumps remain. These tools can maximize convenience for busy and sleepy parents, and ultimately you can make your own decisions about which could actually be helpful during late night feedings and which would just clutter up your kitchen.

Whatever feeding approach you use, digital tracking apps are an easy way to help you (and your pediatrician) track healthy development of baby's duration, amount, and frequency of feeds. When comparing baby tracking apps, consider things like:

- Price: many apps are free, but those with a paywall may include features like Bluetooth connections that seem worthwhile to you.
- Creator: You may not care who creates your apps, but some moms like to know when formula or toy companies have access to shared data from your device through their free app.
- Device compatibility: Do you wear a smartwatch, or keep a tablet charging near baby's favorite rocking chair? Find an app that syncs across your devices so you can conveniently begin and end tracking sessions without disturbing the baby.
- Multi-user tracking: If you have multiple caregivers (another parent, grandparent, nanny, etc.) who watches the baby for substantial amounts of time, it can be very helpful to choose an app that lets other people track data for the baby from their own devices.

Even better, many of these apps also let you track sleep, diapers, and other milestones in one convenient location!

Choosing to Skip Tracking Apps

"As someone who works in educational technology and loves helpful apps and gadgets (I literally used a contraction timer app to track and graph my contractions when in painful labor with both of my children!) I was surprised to find myself choosing to track both of my babies' big four activities (feeding, sleeping, elimination, and enrichment) the 'old fashioned way,' in a notebook.

This started because I was given paper sheets to log feeding and elimination when I was discharged from the hospital, but I continued it because I found the process soothing, like writing in a journal. I also found it easier to have other childcare providers add their notes into the notebook than with the apps I had on my phone.

This was a good reminder for me that even the most helpful technology may not be helpful to everyone. It is fine for moms to choose when they want to do things the 'old fashioned' way, even if it may seem slower or more complicated to others!"

-Amanda Sullivan

Sleeping

Given the importance of sleep for baby's development and mom's mental and physical health, it's a good idea to look into strategies to support your baby's sleep habits. If you're one of the millions of moms suffering from sleep deprivation, you'll be relieved to learn that there are dozens of technologies and online resources aimed at helping babies get to dreamland!

Did You Know?

Despite the fact that newborns need anywhere from 14-18 hours of sleep a day, sleep deprivation is a classic challenge facing new parents. This is because before a baby develops normal day-to-night sleep rhythms (which takes at least three months), their natural sleep and awake cycles last about three hours – much shorter than the normal 24-hour cycle adults are used to!

Those sleepless nights can hit moms even harder than other family members. Dads lose around 13 minutes of nighttime sleep after the birth of a new baby, while moms lose an average of 1 hour of sleep every night. Sleep loss can lead to irritability, anxiety and depression, and higher risk of nighttime accidents and injuries for mom and baby. Stay up-to-date on guidelines from trusted sources like the American Academy of Pediatrics (AAP) to learn safe sleep practices and ways to avoid risks for nighttime suffocation and SIDS (Sudden Infant Death Syndrome).

Many parents find it helpful to start with information and courses available online. Dr. Harvey Karp is a pediatrician and child sleep expert whose book, *The Happiest Baby on the Block*, has inspired dozens of paid infant sleep programs, and includes free video material showing his research-backed approach to helping young infants sleep using the "Five S's." **Swaddling**, cradling baby on their **Side, Shushing, Swinging,** and **Sucking** on a pacifier or nipple. All the steps of this method are based on the idea that newborns fall asleep more quickly in an environment that simulates being inside the womb. Many technologies can also help with this soothing approach.

For example, white noise and baby sound machines are excellent "shushing" tools, and some high-end bassinets are built to rock, vibrate, and sway babies to sleep. With infant sleep, probably the most helpful technology you can get is a tracking app as you get to know baby's unique sleep patterns and average "wake windows" (time spent awake between naps). Since their sleep rhythms are underdeveloped, their schedule is completely unlike yours—and trust us, you aren't as good at estimating how long baby's been awake or asleep as you think, especially when you're exhausted and in the "survival mode" phase of newborn parenting.

As baby gets older and notices his surroundings more, you may find technology useful in developing a bedtime routine. For example, night light projectors can add a soft ambiance by projecting starry skies over baby's ceiling, and many devices even have speakers and Bluetooth capabilities to play baby's favorite lullaby playlist. The popular Hatch night light allows scheduled light and sound modes, and even uses red and green lights to remind older toddlers when it's time to stay in bed or wake up in the morning. It's important to check that any product you use for sleep is aligned with current AAP recommendations to avoid SIDS risks. Heated mattress pads, bouncy swings, and certain in-bed co-sleepers that may be fine for playtime are typically not recommended for infant sleep, and no blankets, pillows, or toys are recommended in baby's crib before they are one year old.

At some point, baby will probably be ready to sleep on their own for a nap or part of the night—whenever that may be! Between our three kids, these authors had two babies who wanted to sleep nowhere but on mom for around a year, and one unicorn baby who loved independent sleep in her crib from early infancy. Be easy on yourself and on your baby as they navigate this big step of sleeping away from you! Some technologies can help children ease into more independent sleep routines, like the screen-free Toniebox and the simple-screen Yoto player. These audio players use toys and cards to allow toddlers to easily choose and play their favorite songs and stories without adult help. They can even play original sound recordings of loved ones reading or singing especially to your child.

The ironic thing about baby sleep is that as soon as you carve out some alone time for mama with baby sleeping peacefully in their crib, you may find that all you can think about is whether they're doing okay in there without you to watch them. Sound and video monitors can really ease a mom's worry, especially with features like being able to talk into a microphone to your child to soothe them if they're fussing, or tilting the camera view if baby moves around the crib. If you already have an in-home virtual assistant like Alexa, you may be surprised to learn that you can also use this like a baby monitor. Keep Alexa in the nursery and an Echo on your end, and say, "Alexa, drop in on the nursery" to hear everything happening in that room. Just be sure to tap the "mute" button on the Echo at your end, so you don't wake the baby.

On-body sleep monitors like the Owlet Sock and HALO Sleep Sure are products that might especially appeal to families of preterm or at-risk babies that need extra tracking, so moms feel like they can act immediately if anything seems off about baby's sleep. These devices track sleep patterns using heart rate, temperature, and movement, and alert parents when baby requires attention. These are a personal preference, since we've known moms who find this kind of tracking crucial to supporting their own mental health after bringing home a baby from a NICU hospital setting where they've been monitored constantly, and others who felt the constant alerts (for example, from the sock falling off the baby and needing to be repositioned) were too stressful for them to use. But for many moms, it can provide much-needed closeness and peace of mind from outside the nursery. One mom we know even loved using her Nanit monitor and associated app to check on her daughter sleeping while she was traveling far from home on business trips.

With all WiFi-enabled monitors (just as with any WiFi-enabled device), you may have heard that people outside your home can hack these video camers. It's extremely distressing, but there are ways to protect your family, such as changing your passwords and factory settings when you first set up a device, using two-factor authentication to login if possible, and adding a VPN (virtual private network) on your router to protect all your WiFi-enabled devices. Read the owner's manual, and choose a device that advertises itself as a "hack proof" monitor. Ultimately, you know yourself and your baby best, so take your instinct into account when shopping for monitors. One benefit of monitors is they can extend far past the early infancy months! We still use ours to check on older kids who can play on their own in their bedrooms (with their knowledge, of course).

Elimination

In early days when baby can't speak for herself, it can be difficult to know how to respond to her cries. Checking for a dirty diaper is easy enough, but how can you tell if it's gas, constipation, diaper rash, or something else? How do you answer when the pediatrician asks if baby is making enough wet and dirty diapers? And where do you check to see if her poop is supposed to be *that* color? Here again, tracking apps will be your best friend and often the easiest way to address questions like this.

As babies grow their poop changes dramatically (even in the first week!) and these changes only become more diverse in the first year as their little digestive and immune systems grow and they start to take in food beyond milk or formula. Poop is one of the best ways to determine how well baby is eating, and "off" colors (like white, black, or red), consistencies, or frequencies in poop are often the first indicator of deeper problems like liver or gallbladder diseases, allergies, infections, and more. Tracking apps usually let you customize beyond simple "wet / dirty / dry" selections to include the color and consistency of infant poop, and some even let you take a picture to share with your pediatrician if you're concerned. Our favorite poop monitor is the PoopMD+ smartphone app. Open to anyone with an infant zero-to-six months, the app is actually part of an ongoing study by Johns Hopkins to grow medical knowledge about infant liver disease. In return for sharing your app data privately with the Johns Hopkins research team, you get access to a free app that uses color recognition software and your smartphone's camera to determine whether the color of your baby's poop is normal or abnormal.

Although poop is a key health indicator, it's important *not* to drive yourself crazy! If something seems off or you have a question, we always recommend bringing it to your pediatrician to get an expert opinion.

Enrichment

You've chosen a tracking app, researched formula, bottles, and breast pumps, and set up your baby's sleep and diaper-changing arrangement. Now it's time for the fun part–play and socialization! If this is your first child, you may wonder what exactly a newborn baby needs for playtime? Come to think of it, what should you expect to *do* with baby all day during your maternity leave? There are dozens of books, resources, and apps to help you answer this question, but for at least the first month or two, expect to be in "survival mode," just making sure you and baby are recovering from the intense physical and emotional toll that birth takes on both of you. It's true that labor is not a medical illness, but don't expect to dive back into your pre-pregnancy way of life so fast!

After the first two months, you'll notice little social interactions and changes–baby's first smile, first eye contact, or first attempt to hold his head up at tummy time–and the growth comes fast and furious from there! Sometimes these changes are so small we wouldn't even notice them if we weren't looking. If you're curious about all the little details of how your child is learning and growing in the first years, you might love an app like Wonder Weeks (thewonderweeks.com). This app uses your baby's due date to offer evidence-based predictions about "developmental leaps," the social, physical, and mental milestones he's exploring. It also offers fun ideas for mom to support baby's development, usually through simple conversations, songs, and games throughout a normal daily routine. We also find so much inspiration from moms on social media who share their creative ideas and life-hacks for babies at specific ages and stages. Several moms we know love creating digital photo albums or YouTube channels shared privately with remote family members to help everyone stay connected and watch the baby grow during these fleeting early years.

You'll probably also notice that babies in this stage of life cry…a lot. And cry. And cry. If you're ever concerned that your baby is inconsolable or might have colic (three+ hours daily of inconsolable crying) you should speak to your pediatrician. But keep reminding yourself that this is probably normal, since babies have literally no other way of communicating right now. We know moms who swear that they learned to differentiate their baby's cries and tell the difference between, for example, a cry for a diaper change and a cry for milk. If you're in the early weeks and still figuring out how to listen for those nuances (or, if you're like us, and you're just waiting for baby to learn to speak so you don't have to interpret a foreign language), you might like the Baby Language app (babylanguage.info). Developed by a linguist who focuses on baby cries, this app claims to tell the difference between five kinds of common baby cries, and gives you a clue of where to start trying to address the problem!

You may be wondering if you need to register for tech toys for your infant? While it's not a bad idea to explore anything you want to invest in for the future, babies in the first six months really need little in the way of technology. They will be perfectly happy chatting, making funny faces, or singing songs with their favorite people, playing with black and white toys and books, and exploring sensory experiences like water, bubbles, and mirrors for several months. If you're looking to distract a baby going through something challenging (two-month shots, anyone?) we have been there, too. However, it can help to know that watching pets, looking out windows, running water, or loud singing/silly sounds with mom can be as distracting and engaging as a phone screen for most young infants. You don't need a tablet or TV for baby's room, and you probably want to avoid any screen time if you can. The one exception to this rule is video chatting with important people in baby's life, especially if it's hard to see them in person. During the COVID-19 pandemic, researchers learned a lot about the impact of social technology time on young children's development. We'll explore what they found more in later chapters, but for now just remember that social interactions (especially where the person on screen talks directly to the baby, and pauses for responses) are always better than passive viewing experiences, and larger laptop or tablet screens may actually be better for baby's tender eyes than small phone screens. With or without technology, try to involve your baby in your conversations as much as possible, and invite others to speak directly and playfully with the baby.

Nursery Cheat Sheet

Keep this cheat sheet handy when deciding about your nursery tech necessities!

TECH NECESSITIES (AND NICE-TO-HAVES) FOR YOUR NURSERY

FEEDING

- Baby Activity Tracker App-Decide if you want an app that supports additional caregivers, and multi-device compatibility (e.g. smart watch)
- Breast Pump, if breastfeeding-See if you can get one covered by your health insurance.
- Bottle Sterilizer, if exclusively bottle feeding-Most moms recommend keeping eight sterilized bottles ready a day!
- Nice to Have: Automated formula mixer or bottle warmer

SLEEPING

- Sound Machine-These can be portable for the stroller or plug-in for home use.
- Light Projector Machine-Soothing, dimmable lights for baby during bedtime rituals, and helpful for late-night wakes. Many are also Bluetooth enabled for bonus music projection.
- Audio or Video Baby Monitor-If choosing a WiFi-enabled monitor, look for "hack proof" brands.
- Nice to Have: Automated rocking/shushing bassinet for baby. These can be pricey, so check rental options and family-to-family donation groups on Facebook.
- Nice to Have: On-body baby sleep monitor for heart rate, breathing, and movement tracking.

ELIMINATION

- Baby Activity Tracker App-We love the PoopMD+ app from Johns Hopkins for specifically tracking healthy diapers.
- Night Light-Important for late-night diaper changes. You can even get diaper changing baskets with night-lights embedded in the handles.
- Nice to Have: Alexa or similar Virtual Assistant tech for the nursery. Helpful for quickly remembering to purchase replacement diapers and wipes. Also doubles as an audio baby monitor.

ENRICHMENT/OTHER TECH TO CONSIDER

- Baby Milestones Tracker App-Helps predict baby's fussy periods and developmental leaps in the first year.
- Digital Thermometers-Get a rectal thermometer for checking baby's internal temperature (most pediatricians only trust rectal readings), and a water thermometer for making sure bathwater isn't too hot or cold for baby's sensitive skin.
- Nice to Have: Smart watch that syncs with baby activity tracker apps, so you can easily log diapers, feeds, and sleep without disturbing the baby.
- Nice to Have: Tablet or laptop for occasional video-chatting with remote loved ones.

This is such a special and intimate time for you and your growing family. Besides finding the right balance of tech tools and other furniture and toys for the nursery, we recommend connecting one-on-one with your baby to foster a strong, secure, and positive bond. Remember that often what's best for your baby is prioritizing whatever you need to be the best mom you can be! Join a mom's group, take your friends up on offers to receive a meal train, and connect with people (via video chat if needed!) who can help you embrace your new identity as a mom (or mom of two, or three…). You'll still be the same person at the end of the tunnel, but there's so much growth to come in the years ahead. Take this time to savor the moment as much as you can, and when things get tricky, remember that this stage of childhood is a season and all seasons eventually pass—and maybe give your family and friends a video call.

PART TWO

Navigating Technology & New Motherhood

CHAPTER 4
A NEW DIGITAL WORLD FOR MOM

"There is no experience in a woman's life that is more impactful, all-encompassing, and life-altering than becoming a mother. The transformation from woman to mother is a psychologically profound experience that both overlaps and is separate from the physical experience of becoming a mother" - *Kate Babetin, Journal of Prenatal & Perinatal Psychology & Health*

The Birth of a Mother

In the midst of the excitement and exhaustion of welcoming a precious bundle of joy, you may have overlooked the birth of someone else: a mother (that's you!). With this new title comes a whole world of new responsibilities, changes to your sleep and self-care routines, along with changes to your social and professional life. All of these big changes, paired with hormonal shifts or other physical changes you may experience, are enough to take a toll mentally and emotionally.

The process of becoming a mother, which anthropologists call "matrescence," has been largely unexplored in the medical and scientific research community. Instead, medical care and research in the U.S. has been more focused on supporting the wellness of new infants and children, and the role mothers play in that, than exploring the major identities transition women undergo. As women today are navigating their new identities as a moms, along with the emotional and mental hurdles that go along with it, they often turn to the Internet and social media for reassurance, companionship, guidance, information, and advice. The Internet also provides a space for women to explore portraying their new identity publicly, through sharing images, videos, and text about their experiences in motherhood. New moms may look for "mommy accounts" to follow on Instagram or TikTok, blogs with information on navigating the baby blues, or online groups to ask questions or vent their exhaustion and struggles anonymously to an audience who will "get" them. Both of us have used (and continue to use) social media and the Internet in these ways!

If you, like us, engage with "mommy culture" on the Internet and social media, you might wonder how these online interactions may impact the postpartum experience for new moms or what the long-term impact these interactions have on moms' self-esteem and confidence later in their motherhood journeys. This chapter will explore everything from mommy Facebook groups, platforms like Meetup for in-person mom gatherings, and new motherhood "matchmaking" apps like Peanut, and the influence of these spaces on moms. We will also provide strategies for navigating and using these digital resources to form a positive community of moms.

But first, a trigger warning: This chapter will explore some themes and ideas around postpartum depression, the baby blues, depression, and maternal mental health. We explore research around these themes as they relate to the use of digital technology. Please reach out to your doctor for guidance if you think that you or a loved one is suffering from postpartum depression.

Did You Know?

Recent research from the Centers for Disease Control (CDC) has found that about 1 in 8 women experience symptoms of postpartum depression. Effective depression treatment can include a combination of medication therapy, counseling, and referrals. The first step in treatment is talking to your healthcare provider.

Postpartum depression is depression that occurs after having a baby. Feelings of postpartum depression are more intense and last longer than those of "baby blues," a term often used to describe the worry, sadness, and tiredness many women experience after having a baby. The CDC lists these symptoms of postpartum depression:

- Crying more often than usual.
- Feelings of anger.
- Withdrawing from loved ones.
- Feeling distant from your baby.
- Worrying or feeling overly anxious.
- Thinking about hurting yourself or your baby.
- Doubting your ability to care for your baby.

If you or someone you know is struggling with postpartum depression and needs help, reach out to these resources:

- Maternal Mental Health Hotline: 1-833-9-HELP4MOMS (via text or a phone call)
- National Suicide Prevention Lifeline: 1-800-273-TALK (8255)

Call 911 for an Emergency: If you are thinking about harming yourself or your baby, or if you are concerned about someone, call now.

Embracing a New Digital Identity

The majority of parents in the U.S. use some kind of social media, and mothers (particularly young mothers) are among the most active social media users out there. In this section, we explore benefits and challenges facing mothers on social media. We'll present research showing that social platforms are no replacement for in-person and real-life support networks. We find that most moms already know this, so you might ask, why is social media so popular among parents anyway. This could be because social media allows women to experiment with how they share and express their newfound sense of identity as a mother. This journey of moms' identity exploration online typically starts with something as simple as posting pictures during a pregnancy or adoption journey or baby milestone pictures during their first year at home. Research by Schoppe-Sullivan and colleagues found that typical moms report first uploading a photo of their infant to Facebook within one week of their child's birth and 80% of mothers who had ever uploaded a photo of their child reported that they had featured their child in their profile picture. Those mothers who used their child's image in their own profile photos tended to show stronger identification with their role as a mother than women who didn't.

What does this all say about the identity of new moms? In an article posted on *Science Daily*, *Schoppe-Sullivan*, the lead researcher of the study described above, explained that "what these mothers are saying is that my child is central to my identity, at least right now. That's really telling," Schoppe-Sullivan said.

Aside from sharing photos of their child, research also shows that moms have a long list of reasons to enjoy using social media, all of which may help to formulate their shifting identity. For example, social media offers moms a platform to share content related to their children with other parents and to follow the postings of others who have similar views on parenting for tips and strategies. For moms who are still unsure how they feel about certain parenting "hot topics" like nursing or sleep training, social media offers women a chance to hear and explore different viewpoints in a relatively safe space—often as a "fly on the wall" without having to voice an opinion the way you may need to during in-person interactions. Social media apps like Instagram offer unique opportunities for instant communication, connection, and visual information—even in the middle of the night when you are nursing a sleepless baby and it might feel like no one else in the world is awake!

As social media and the digital world becomes an increasing part of how women develop their new sense of identity as a mother, it is important to look at the impacts of social media usage on mothers' self-esteem, confidence, and sense of community. The following section dives into the impact of digital experiences during motherhood and their positive and negative influences on maternal mental health and emotions.

Did You Know?

Do you need even more proof that having a baby dramatically changes a woman's mental state? A 2016 study published in Nature Neuroscience revealed that during pregnancy, and for at least two years postpartum, a mom's brain undergoes significant remodeling. Areas associated with social cognition and empathy change so dramatically that a computer algorithm could accurately predict who had been pregnant simply based on brain scans!

How and why does this happen? Researchers theorize that moms' brains are becoming more specialized to adapt to motherhood and better respond to the needs of their babies.

So the next time you feel you are losing your mind adapting to new motherhood... go easy on yourself! Your brain is doing a lot right now.

Digital Culture and Maternal Mental Health

The Benefits and Challenges of Motherhood and Social Media

We asked moms to tell us about the positives and negatives they have experienced with social media since becoming parents. Here is what they had to say:

The Good...

"Quality follows on Instagram and supportive Facebook groups have definitely informed [my] decisions and pointed me towards resources, usually books, that have been instrumental."

"It was really helpful to not feel as alone in the first couple months [after having a baby] when other people posted questions and I was able to read through replies. I'm not much of a poster, but reading by myself helped so much."

"The support - people letting you know you're not alone, and people sharing their struggles."

"Realizing that many other moms are experiencing similar worries and challenges. Feeling less alone."

"Following doctors and nurses who can't give medical advice but show you how to do things you have never done was super helpful."

The Bad...

"Mis-information, unsafe information, and if you're vulnerable to it, near-constant judgment."

"Comparing yourself/your child to others."

"Other people's 'best of reels' can make me feel like less of a mother when I am not doing extra special things, or out enjoying the nice weather, etc."

"You only see the positive aspects of motherhood."

Motherhood is stressful. Whatever your views on parenting, that might be something we can all agree on. From a research perspective, we have known about the stressful nature of new motherhood for quite some time. Studies of new mothers in the 1980s showed that their stress levels increased from the time they were six months pregnant to at least eight months after the birth of their baby, and new mothers generally had higher postpartum stress levels than new fathers. Flash forward to today: not much has changed about how stressful parenting is! Moms today, more than dads, say parenting is tiring and stressful and that they feel judged by certain groups for how they parent. Can anyone else relate?

When women do not receive the support and help they need, this stress may impact how moms transition into parenthood and can even lead to other emotional and psychological issues. So what gets moms through this deeply stressful period in their lives? Positive relationships and support. Research has shown that for new mothers in particular, social support is a critical aspect of successfully navigating the challenging transition to parenthood. While this may have always been part of how new moms transition into parenthood, what has shifted in the digital age is *how* and *where* women are seeking this support. Google searches, Facebook groups, TikTok accounts, Instagram influencers, and more are now at the fingertips of new moms and a huge way that moms seek the support and information they need.

So what exactly *is* the impact of all these new websites, apps, and social media on moms' mental health and emotional well-being? Well, the answer is a bit mixed. One of the most notable benefits of social media for moms today is the ability to find a sense of community and support, even if you do not have that support in your "offscreen" life. Moms can use social media to connect with women and learn through sharing personal experiences. Moms can also use social media to find resources and information. Many moms use mommy-focused groups on Facebook, for example, to ask questions and make informed decisions about their children's education and healthcare. Moms are also able to learn new information in digestible ways that are easy to consume while multitasking, such as through short videos, online classes and webinars, and more. All of this can lead to an increased sense of happiness and confidence for moms in decision-making and purchases around baby tools and parenting choices.

Social media also allows moms to have an outlet to share their own personal experiences in parenting through updates, photos, and videos. Posting these milestones can help friends and family feel connected and close even if they live very far away. This was something that became even more important during the start of the COVID-19 pandemic, when many families could not be together in person, and continues to be one of the main benefits of social media use today.

On the other hand, negative interactions online may make moms feel more isolated and increase their stress. For example, a study conducted by professors Nataria Joseph, Theresa de los

Santos, and Lauren Amaro, in 2022 at Pepperdine University found that the more time women spend in mom-focused social media groups, the higher their cortisol (a.k.a. stress levels) are. Researchers found that the spikes in cortisol were likely due to "negative interactions with other moms on social networking sites and more time spent with these interactions." This study is pivotal as one of the first to find that negative experiences, including social-cognitive and emotional reactions to technology-mediated social exposure, can have a measurable impact on daily cortisol levels.

Part of the reason social media interactions can have such a stressful impact on moms is that we often use social media to get the affirmation we need to feel good about ourselves as parents. Researcher Sarah Schoppe-Sullivan explained to *Science Daily* that "if a mother is posting on Facebook to get affirmation that she's doing a good job and doesn't get all the 'likes' and positive comments she expects, that could be a problem. She may end up feeling worse." Schoppe-Sullivan's research concluded that mothers who posted more on Facebook reported more depressive symptoms after nine months of parenthood than other moms.

Experiences on social media don't just cause stress and depressive symptoms; they can also cause mothers to feel envy and anxiety, especially when mothers are consuming overly idealized and unrealistic depictions of motherhood. In a 2022 online experiment with 464 new US-based mothers (i.e., mothers with a child 3 years old or younger) that was published in the journal *Computers in Human Behavior*, the researchers exposed participants to Instagram posts portraying motherhood that varied in idealization (i.e., idealized vs. non-idealized portrayal) and source (i.e., posted by a mommy influencer vs. an everyday mother). Regardless of the source, whether it was a regular mom or mommy influencer, the idealized posts cause significantly higher levels of envy and state anxiety, which may be detrimental to mothers' overall mental health. In an article published in *Motherly* online, Dr. Ciera Kilpatrick, one of the study's authors, explained that, "idealized portrayals of motherhood aren't new to the media. In the '70s and '80s, idealized portrayals of motherhood were all over magazines in the form of 'Celebrity Mom Profiles.'" But now, they're all over social media. And while it was celebrities glamorizing motherhood in those magazines, anyone can glamorize motherhood today because of social media." When consuming overly glamorized and staged media about motherhood every time we scroll on our phones, it is no surprise that many moms feel bad about their own experiences parenting or feel like they are not doing enough.

This may sound like there are a lot of negative outcomes associated with engaging with mommy culture on social media today. But keep in mind that the main takeaway here is NOT that using social media apps like Instagram or Facebook is necessarily harmful or that these sites are "bad" for moms. Looking at the accumulated research in this area tells us that these sites have a lot of benefits, but they may not be the most effective platform for women to gain external validation of themselves as moms. Donna Moore, an expert in maternal Internet use at City University in London, explained in an article published in Vice (online) that, "the Internet is a double-edged sword for new mothers. There are dangers with overuse of social media and relying too heavily on validating your maternal identity through your interactions online."

Okay, so how can moms today get the validation that we all need? It may be helpful to seek this validation offline, through conversations and interactions with a spouse, partner, close friends, and family members. Even self-reflection on interactions with our children can also help to provide this validation in a meaningful way. Don't worry, we will circle back to more actionable tips and advice at the end of this chapter!

Mommy "Matchmaking" Apps

We now know the key takeaway from the research is that we moms need to seek validation, support, and connection, when possible, through offline interactions. This might come from our partners, friends, or family members. So why does it feel like those interactions don't always scratch the itch we get from online mommy culture and digital connections? Probably in part because so much of what women are seeking online is not just support and validation: it is support and validation from **other moms in the same season of motherhood** that we are in. That means it doesn't matter if we have the most supportive husband, mother-in-law, or childless bestie in the world. That support is AMAZING, don't get us wrong, but it just isn't the same as getting it from another mom going through the same stage of motherhood we are in, whether it is the sleepless newborn stage, the terrible twos, or the unpredictable teen years: we want someone going through the thick of it alongside us.

But what happens if we don't have those other moms in our daily "offline" lives? Online mommy culture can help fill that void—as long as we are careful to avoid the dangers discussed in the previous section. Digital technology and apps can also help us bridge the gap into "real-life" friendships and community. We've all heard about dating apps like Tinder, where you swipe left or right to make a match with a potential romantic partner. But did you know that there is a host of "mommy matchmaking apps" designed to help women make new friends who are also moms? If you are a new (or veteran!) mom who is still looking to grow your "village" or if you are simply looking for new people to meet up with for playdates for your kids, one of these mommy apps might work for you:

- **Peanut:** This popular and easy platform works in a similar way to Tinder, offering a swipe up/down "wave" function. The goal of Peanut is to provide a space where women connect with women who are at a similar stage in life–from fertility, pregnancy and motherhood, through to menopause.
- **MomCo:** This app lets you locate moms that live close by with similar-aged children and common interests.
- **Hello Mamas:** This app lets you meet local mom friends, get to know your community, receive community-wide alerts, and more!

Beyond these apps, many local Facebook groups for moms are designed to facilitate a connection that goes beyond the screen with scheduling playdates, parties, moms' nights out, and more.

Finding Friends Online

Real Mom Reflection

As a military spouse, I have learned over the years to adapt to moving frequently. The first time I moved across the country as a milspouse was when I was nearly 9 months pregnant with my first child. During this time, I transitioned my career to working fully remotely and left my friends and support system behind. I think I underestimated the ease with which I would find friends in my new city. Every other time I moved as an adult was for school or a job, so I met people with similar interests almost immediately. But this time was different, and I felt utterly alone and disconnected. I went to baby music classes and baby gym groups every week and my son and I were always very social with playdates, but I never found my own friends. I felt extremely lonely and isolated, but hid it expertly under my busy baby class agenda and major professional milestones. Only those closest to me knew how painful those first years of motherhood were for me.

When we moved again, I made a promise to myself that things would be different. Shortly after my second child was born, I signed up for an account on Peanut - a mommy friendship and networking app with a similar user experience to Tinder. I swiped and talked to many moms via text and eventually met one mom in person at a playground. It was like an awkward first date but we actually hit it off and became very good family friends over the next couple of years. I am forever grateful that we matched on Peanut and that, although we no longer live in the same city, she and her family will always be a part of my life!

Here I am in another new duty station, and back on Peanut and Facebook. **I know that I may not meet another bestie, but I am thankful these technologies exist to help moms like me find friendship and connection when we need it most.**

-Amanda Sullivan

Social Media Do's and Don'ts for Moms

We promised earlier in this chapter to share actionable tips and advice based on all of the research and information shared in this chapter (and we know it was a lot to take in!). If you are wondering how to make sense of all the mixed negative and positive findings about using social media, use the following "Do's" and "Don'ts" to help guide you for a few weeks. Then, check in with how you are feeling and come up with your own personal guidelines for engaging with social media in a way that makes you feel happy, connected, and confident!

Do's:

☑ **Have fun on social media:** First and foremost, use social media for fun! Use social media if it makes you smile and you are genuinely having fun using it. This might be consuming silly videos, editing and sharing your own content, or getting joy by following updates from your friends and family far away.

☑ **Share updates you're comfortable with:** Using social media to share updates and milestones from your parenting journey can help you stay connected with friends and family who are far away! Deciding on content and photos to share can also be a positive outlet for you as you are developing your new identity as a mom. Remember to only share content you feel comfortable sharing, and that you can always check your privacy settings on these different apps and sites if you aren't sure who is able to view the content you share.

☑ **Follow uplifting accounts:** Whether they are the accounts of influencers or people in your offscreen lives, choose to follow accounts that are supportive, uplifting, relatable, or otherwise make you feel good. This does not mean that they can only share happy content, in fact, seeing moms share "#momfails" and their bad days can actually be very grounding and relatable! We just mean that you should choose to follow accounts that are welcoming, inclusive, and make you feel like you belong.

☑ **Take information with a grain of salt:** It is okay to learn from "mommy influencers" and from the personal experiences of other moms online! Just take it all with a grain of salt and remember to contact a professional if you need real guidance about your health, mental health, or your child's well-being

☑ **Seek connection and community**: Use social media to find a sense of community and connection with other moms.

☑ **Take a break if you need it:** Just because you enjoy social media use, doesn't mean you need to use it all the time. Think about finding ways to limit your social media use and be mindful of how much time you spend scrolling through your feeds (sometimes we don't even realize an hour or two disappearing in the rabbit hole of scrolling—especially at night!). If you find that any of your experiences online are starting to cause you stress or anxiety, allow yourself to take a break and focus on your own well-being.

Don'ts

☒ **Don't use social media for validation:** If you are looking to social media to feel good about yourself or for reassurance that you are a good mother, unfortunately, this may lead to negative feelings in the end. Rely on other sources (including your own personal observations on your interactions with your children!) to validate yourself as a mother. Also, remember to find validation through your accomplishments in other domains outside of motherhood! Yes, we know this book is all about motherhood, but motherhood is NOT the only thing that defines you. Try to find validation in your successes in your professional life, through hobbies, or in your role as a friend, spouse, or family member.

◪ **Don't let social media impact your sleep or other healthy routines:** If you find yourself staying up all night scrolling, it may be time to cut back on your social media use. Social media should add to the fun of your day, not take over your day (or night, for that matter).

◪ **Don't follow accounts that make you feel bad about yourself:** Whether they are the accounts of influencers, celebrities, or simply friends and family members, you should not follow any accounts that make you feel bad about yourself or your approach to parenting. If there are any accounts in your feed that make you feel envious, anxious, or in any way negative just tap "unfollow" and say, "Thank you, next!"

CHAPTER 5

AGES ZERO-TO-TWO: STAYING PRESENT AND LIMITING SCREENS

"In 1970, the average age at which a child began to watch television regularly was four years. Today, it is four months." - Erika Christakis, author of *The Importance of Being Little: What Preschoolers Really Need from Grownups*

Unless you're extremely disciplined and have a very strict no-screen time policy with your babies, you have probably found yourself at some point in your parenting journey using screens to distract or entertain your toddler. In fact, 90% of American moms use some kind of screen habit with their baby before they turn two years old. Maybe at your house, that means letting the baby play with your phone so you can finish a quick errand, enjoying dinner around the TV with an infant or toddler in your lap, or making daily video calls with your baby to chat with distant relatives. You may be wondering how much your baby is paying attention and impacted by your screen and technology time—and what you can do to maintain your sanity if the baby's screen time is part of mama's me-time! This chapter summarizes what the research currently says about the effect of screen time specifically for babies under age two and shares some ideas for balancing the recommendations with real life. But first, we asked other moms about their challenges with staying present and limiting screens. See what they had to say in the Real Mom Reflections below!

Limiting Screens with Babies & Toddlers

We asked moms about their experiences with moderating screen time, and here is what they had to say:

"I know the recommended amount is zero screen time for my 8-month-old, but I feel like everything in moderation."

"I think at her age no screen time is appropriate; however, I think that's easier to implement with one child in a two-parent household than for single parents or two-parent families with multiple children."

"Sometimes we disagree about having it [television] on in the background. I try to get her away from the TV if it is on, but my partner is not always as adamant. Our thoughts and conversations about it continue to evolve as she has started paying more attention to screens in recent months, and will continue as she gets older."

Generation Alpha: The Babies Born with Screens

Generation Alpha is the term used to describe babies and young children born between 2010 and the mid-2020s. The beginning of this generation shares another important birthday: the first iPad hit the market in 2010 (iPhones were only a few years before that in 2007). This means that Gen Alpha is the first cohort of young children to have grown up in a world where screens and technology are truly ubiquitous. In the US, there is almost nowhere these kids can go where they can't find access to a smartphone or computer and the internet. On top of this, research published in the Journal of the American Medical Association (JAMA) Pediatrics in 2023 confirms that Gen Alpha babies had a *lot* of unintended screen exposure during the global COVID-19 pandemic of 2020-2023. While the evidence is new, and only time will tell the true effects of this social experiment, researchers have learned a lot about how screens can and can't support young children's learning and development in the last decade.

Interactive screens are a relatively new phenomenon for parents to deal with. And yet, you may not remember a time when you didn't have a touchscreen handy at almost all moments of your life. If you're like a lot of moms we know, you might have seen young kids playing quietly with smart devices at restaurants or coffee shops while you were pregnant, and secretly promised yourself you would *never* be that kind of parent…only to fast forward a few years and find yourself passing back your phone to your screaming toddler in the car, or putting on a YouTube video for your baby after a 5am wake-up. We understand your struggle, and we are not here to judge! Moms get enough of that from the rest of the world (and from ourselves). But you may wonder about the impact of your habits: How much screen time is too much? Does the size of the screen matter? Are video chats, TV, and interactive apps all the same kind of screen time? Let's dive into some common questions about screen time and consider what the evidence says, and what we still don't fully understand.

What's the Problem with Screens?

Recent research is clear–and the APA and WHO agree–that ***there are no benefits*** to ***introducing screens before two years old.*** Specifically, the APA recommends no screens before 18 months, and limiting screen time to *occasional* co-viewing of "high quality programming" (think PBS) with active adult interaction from 18 to 24 months. Contrast this with recent research from medical journals like Scientific Reports (owned by Nature) and JAMA Pediatrics that children aged zero-to-two years were averaging three hours a day of screen time in 2019, and that number jumped even higher during the pandemic.

Why are doctors and educators worried about this trend? The American Academy of Pediatrics explains that besides the harmful effects of blue light (light created by screens) on developing eyes, *any* time spent using screens is taking away from other activities like playing outside, exploring toys, or talking and reading books with adults. Time lost to these other critical experiences can be truly harmful for babies. Several studies have found connections between early screen use and later addictive behavior, poor sleep quality, and a higher risk of obesity. Some of the best-designed studies to come out in the last few years have conclusively linked two+ daily screen hours in zero-to-two-year-old children to:

(1) Lower behavioral, cognitive, and social development milestones at three years old;

(2) More attention problems at seven years old; and

(3) Lower language acquisition and cognitive test performance at eight years old. Even "educational" screen time is not great for babies and toddlers.

PBS shows like Sesame Street and Daniel Tiger's Neighborhood, which have been shown to be beneficial for three-to-five-year-olds, don't have the same effect with babies and toddlers, who really don't understand what they're seeing and can't retain information from a TV show or app. Furthermore, "background" screen use, such as leaving a TV on in the room while playing or doing something else, seems to be just as distracting as active viewing for babies (who can't filter out sensory information), and is associated with lower toy use and disrupted free play.

Video Chat: A Screen Time Silver Lining

The research on screens for babies sounds scary, but there are some important positive things to remember before you panic! One thing we learned during the pandemic is that not all screen time is created equal. Video chatting is a major exception to the "all screens are bad" opinion of the APA. As it turns out, the effect of screen-based real-time communication with loved ones is more or less fine for babies, and is even beneficial for moms! That's because video chats involve more back-and-forth talking, facial expressions, and physical interactions than TV shows, so they don't take away from social adult time that children need for important emotional development. Importantly, they also don't cut into the number of words spoken around the baby, which research confirms is a primary challenge of screen time for infants and toddlers. There is even some early evidence that screen-free phone calls may be *more* confusing for babies than video calls, since they don't have the visual cue of facial expressions and gestures to interact with the caller.

Did You Know?

Video chatting is fine for babies, and can even be good for moms!

The early months and years of childhood can be tough, especially for first-time moms. Any support helps to build a mom's endurance and confidence to face the tasks of daily parenting, and that includes emotional support from a distance. Research shows that **moms who stay in touch with their families and friends over the phone show a lower risk of postpartum depression, anxiety, and suicide.** Their babies also have overall more positive health trends than moms who feel isolated.

As a fun bonus, **social interactions during a video call can even support social bonds with the caller in real life!** Babies recognize and prefer people with whom they've had regular video chats. So there's no need to feel guilty about daily video chats with Grandma!

Strategies for Parenting Around Screens

What can you do if your family has already built a strong screen time habit? In a 2018 interview about how parents can help babies cope with the constant presence of screens in their lives, digital education reporter Anja Kamenetz said, "We have to update our strategies, but we don't have to update our values. If we have an understanding that it's good to connect in 3D space offline, we're going to have to protect those spaces". It's true that there is no replacement for young children having real-life, social, and play interactions, but it's also true that those interactions don't always happen naturally in the average American household. As with feeding, sleep, and everything else in a young child's life, babies depend completely on mom and their other caregivers to help them regulate their interactions with technology. And just like in those other areas of your life, you have some important parenting tools in your toolkit when it comes to baby screen time! Regular routines, core family values, and co-viewing or co-play with your child can all help address technology habits you want to change, and build new ones that better serve your family.

When we think about where screens come up in our children's lives, it's helpful to think about *why* we may rely on screens in the first place. Below are some common baby and toddler parenting goals, and some evidence-based strategies for meeting the same intentions while limiting screens. Keep in mind that the best recommendations still call for *no* screen time, but if that feels like a completely unattainable lifestyle change, then these "second-best" ideas are for you.

Education—If you've read this far into the chapter and you've been using high-quality educational TV shows or apps with your zero-to-two-year-old, you might now be convinced to save it for their 3rd birthday. Instead, you can offer classic baby toys like soft books or board books (just look out for teething babies who might try to eat them), simple shape/color/texture toys, or non-breakable mirrors. If skill practice and academic achievement are your ultimate goals, rest assured that the most educational experience on earth for a little baby is lots and lots of social interaction with their favorite adults!

Fidget toy—If your baby is used to you handing them a phone, TV remote, or other technology as a teether or plaything, you can take comfort that extended sight-and-sound engagement with an actual screen seems to be the cause of most negative outcomes associated with screen time. If they're simply using the tool as a fidget toy (rather than watching or engaging with content), you can probably distract them enough to swap it out for a more interesting fidget. Or, just let them keep playing with the tool if it doesn't bother you. Keep in mind the safety checks of allowing a child to play with something not designed for young children's use:

- Avoid remotes or technology with small pieces, batteries, or buttons that could come loose and become choking hazards. **Swallowing batteries is a serious health risk for children that can result in internal acid burns and even death.**

- Disinfect phones, which carry a huge amount of bacteria from being handled all day.

- Check that anything they are putting in their mouths (especially anything with metallic paint) doesn't contain lead. You can purchase simple and affordable lead testing swabs from a hardware store to confirm this.

Co-play with a caregiver—If you and baby love to play together on your device or while watching TV, being very interactive is the easiest way to keep these exchanges enriching for your child. Taking selfies (or "us-ies") where the baby can see your faces, using the phone camera like a mirror, choosing music together, and singing/playing along with interactive videos are all better alternatives than passively watching videos or silently playing apps together (remember, children don't understand apps until about age 3). Co-viewing TV with lots of talking and face-to-face interaction is one of the best ways to mitigate the negative effects of screen time. Try to keep screen play brief (less than ten minutes) and use your interactions to spark off-screen play that is likely more engaging than the device.

Occasional on-the-go distraction—Often, a child's first exposure to intentional screen time is during an overwhelming experience that is unavoidable in the real world, but simply not that great for little kids. We can definitely relate when a mom pops on a YouTube video while her toddler is sitting on her lap at a busy airport, bored to tears in a waiting room, or panicking during a shot at the doctor's office. The truth is, these interactions are short and occasional enough that the level of screen time won't harm your baby. If you feel that it's truly the thing that will calm them down best in that moment, do what works for you both. However, sometimes in these situations (which are usually fairly brief and infrequent anyway) you might find an opportunity to reconnect and be present in the moment. Consider these moments through the eyes of your child. Confusing, frustrating, or scary experiences are often quite stimulating enough without an overstimulating screen competing for their attention. Instead of avoiding these overwhelming feelings, you might take the chance to reassure your child through your calm and constant presence during scary moments, and reconnect through your delight in playing and talking with them during boring ones. Don't hold yourself to an impossible standard—some situations truly require pulling out all your tricks, even if that includes a quick screen distraction. But the more you practice and instill trust in your baby that you are there for them during hectic moments, the more they will be able to self-regulate when they encounter trying times on their own.

- o   For other great (non-tech) on-the-go distractions, it's always a great idea to throw some travel toys, keychain fidgets, or cloth picture books into your diaper bag.
- o   Keep in mind that *occasional* or *extreme* circumstances are not the same as regular experiences. Everyday events like daily commutes, trips to the grocery store, and visits to a restaurant might feel like moments to reach for a device, but falling into a screen-time habit during these everyday moments can add up to the high levels of screen exposure that research warns about. Do your best to avoid that impulse, keeping in mind how easily these "just this once" moments can turn into long-term habits and practices as your baby grows older and expects screens in daily-life situations.

Hands-free babysitter–This is one of the most common scenarios where parents feel dependent on screens. TV and touch screens can offer quick and easy baby distraction that lets mom get a few minutes to finish a chore, send an email, or just enjoy a cup of coffee. While we can't argue with the effectiveness of this occasional distraction, we know it's not the best way for babies to spend time, and certainly not a one to two hour-a-day habit you want to build in your home. Furthermore, we recommend never leaving babies unsupervised with screens on. So how can you get a moment to yourself without a baby on your hip, or a toddler getting into the trash can?

What screens offer is a baby's constrained attention, so they literally forget to look for you and try to explore the world. Instead of constraining them *mentally*, why not try a safe *physical* constraint? Both of these authors are big fans of baby wearing for soothing the baby and freeing your hands! If you want more freedom of movement, many high chairs now offer newborn inserts to lie a baby down at taller height, which offers them a fun adult-level view of the kitchen or living room. Offering children a safe place where they can't get hurt and can freely explore, such as a playpen or a corner of a room that is sectioned off with baby gates, is an excellent way to reap the benefits of playful exploration without having to hold your baby constantly. Younger babies can use this space for tummy time, and toddlers can use it to practice their new crawling, walking, and sensory skills. Although it's important that the space is safe even without adult supervision, you'll want to set this up in an area where you can still see and hear each other, so they know you're nearby and you can keep an eye on their activity. If babies resist this freedom initially, take a few minutes every day (maybe just a few seconds to start!) during a calm moment in their routine to help them get used to enjoying their chair or play space without you holding or touching them. You may be surprised how quickly they learn to enjoy their newfound independence!

○   If you choose to set up a play space, consider Janet Lansbury's suggestions to create a space that says "yes", where any activity is ok, and there is no way to get hurt accidentally even if no adult is watching. Check furniture sturdiness, choking hazards, sharp/heavy toys, and electrical outlets for safety.

In-person family time—Before baby, you may have spent lots of family time around the TV or computer as part of your family bonding time. If enjoying favorite sports teams, game shows, or other high quality, family-friendly content is a part of your at-home culture, we certainly don't advise changing your values for the sake of a new family member. Remember that the most damaging thing about screen time isn't necessarily screens themselves, but that they take time away from face-to-face interaction. Try whenever possible to arrange screen time during times in baby's schedule when they wouldn't necessarily have face-to-face time anyway, such as naps. Screen time at a normal volume or with headphones while the baby is sleeping on you is completely fine! Just be sensitive to their needs if the TV tends to wake them up.

Counterintuitively, if you are committed to shared screen time, we actually recommend a family-friendly show that everyone likes rather than something like a Baby Einstein video, since adults are more likely to talk and co-view during a show they are actively watching. Family movie nights or sports games with the neighbors offer plenty of social interaction opportunities. Talk frequently with the baby about what's happening around them and what others in the room are doing, offer alternative toys or play opportunities, and use commercials to play with your baby on the floor, bounce and carry them around the house, or simply turn the TV off for a few minutes. No matter how interactive you are, keep in mind that babies' attention spans are much shorter than adults', and you may need to break up longer viewing sessions into five to ten minute chunks (or just remove baby from the room). If you notice that it's hard for babies to tear their attention from the screen, this is a sign of overstimulation and they need a sensory break with a caregiver away from the screen. If you're lucky enough to have a family member who doesn't mind missing the movie, invite them to hang out with the baby. You can still enjoy a group gathering before or after the showing, and baby will have a blast during special one-on-one time with their babysitting adult!

    o   If you're concerned about the amount of screen time that your family shares overall, a great antidote is playing outside! This can be a fun way to connect off-screen play with on-screen content, for example, by playing with a ball after watching a soccer match. Babies are never too young to spend time outside each day, and toddlers can learn so much from playing in a park, yard, or even a playpen near a window. Outside time won't offset language delays in babies with heavy screen use (regular one+ hour a day), but research shows that daily or almost daily time spent playing outside can help improve socialization and later life skills in babies with high screen exposure.

 o No matter what strategies you choose, remember to keep stints of passive screen use to a moderate amount, if you can't avoid it completely.

Remote family time–This is an easy one: go ahead and video chat! These live interactions are not harmful for babies, and are very good for mom's mental health. Use larger screens when possible to avoid straining the baby's developing eyes, and hold the device at least 2 feet away from their face. Young babies may still be slightly confused by this type of interaction (our kids liked to check behind the phone to see where the caller was hiding), so try to keep calls highly interactive. Involve them in the conversation and invite callers to speak to kids directly. You can also have callers read books, eat food, or use toys that you also have on your end of the call, so baby can experience shared interactions with all of their senses. Invite callers to send kisses or hugs that you act out on your end. Use a special greeting every time you call a specific person, and then continue this greeting when they visit in real life, so baby can build on that familiarity. Feel free to use video chats frequently! Just remember that brief calls are more respectful of baby's developing attention span than lengthy chat sessions.

At the end of this chapter, we've included a tool to help you reflect on your own parenting habits in these common scenarios, consider where screens might show up in your zero-to-two-year-old's day, and what other strategies you can use in those moments.

Practice Being a Role Model

Screens are more commonplace today than ever before. Sometimes the hardest habit for families to build isn't about providing boundaries around screen access directly to children, but modeling how adults can engage (or disengage) with screens in a healthy way. Taking the first step to limiting screen time around kids could involve putting down the phone and turning off the screens when you're with them. Revisit Chapter 2 for tips and tools to examine your own technology habits.

Navigating Screen Time in Infancy

Real Mom Reflection

Before I became a mom, I knew all the research behind screen time and the negative consequences that can happen with overexposure. After my baby was born, our family fell into a comfortable routine that involved no screen time during the day – but we struggled with finding a way to relax in the evenings after work, when the baby's bedtime was still very late and unpredictable (especially since our favorite ways to decompress involve shared TV and video games). My husband and I were tired, cranky, and felt like we didn't have any time together if one of us wanted to watch a show while the other parent stayed with the baby in a separate room. When loving family members would offer to babysit so we could leave the house and enjoy some time together as a couple, we would return to stories of the 'educational' TV shows they watched with the baby while we were out. I felt like the screen police, but I didn't want family harmony to suffer!

My husband and I talked, and we agreed that in the big picture, our baby wasn't anywhere close to the 1-2 hours of daily screen time that is associated with the worst learning outcomes. We also agreed that a strict "no screen time" policy wasn't supporting our family value of togetherness during relaxation time. Instead, we agreed on a plan to get a bit of shared screen time through the week:

-Co-viewing nights: A few nights a week, our daughter would stay near us in her playpen for short TV shows that we were likely to talk through (usually 20 minutes of the game show Jeopardy!).

-Low-distraction nights: One or two nights a week, we would quietly watch 20-30 minutes of recorded shows or live sports on very low volume and with closed-captions on while she nursed or slept in my arms on the couch (this was in the early days when she wasn't easily distracted while nursing!).

-TV-free nights: The rest of the week, we wouldn't turn on the TV until after bedtime when she was fast asleep. Sometimes this meant no TV at all, if she was having a fussy night. As she got older and developed a regular bedtime, this became the most common routine for us.

Our routine still involved more screen time than some families would feel comfortable with, but having a plan that was realistic for our household helped me feel less guilty, and allowed all of us to enjoy some downtime together! -Amanda Strawhacker

Even if you've made a personal decision to limit screen time around your child, you may find that it's hard to change your entire household's habits. Remember that just as your baby has a right to a screen-limited childhood, your older loved ones have a right to meet their legitimate needs for relaxation, social connection, education…and often, those needs can be met without screen time. It's important to enter into conversations about family screen time as collaborators and teammates. Initiate open conversations with other caregivers and older children in your home about shared family values and technology limits specifically when your zero-to-two-year-old baby is present. Avoid blaming or calling out habits that you don't personally agree with. If your family decides on a habit that definitely needs to change, try to commit to a small, manageable start that works for everyone. Every family is a work in progress, and whether you embrace screens, reject them completely, or seek a balance, your decisions should help connect you and your family around your shared values and screen time agreements. In Chapter 9, we'll revisit the topic of healthy screen habits for the whole family. In the meantime, any small changes to lower screen access in your home for your babies under age 2 is a step in the right direction that you can feel proud of!

Screen Time Swaps: Parenting Your Zero-to-Two-Year-Old with(out) Screens

In this table, look at each parenting goal that could cause you to reach for a device to hand your child. Use the second column to check or write in any screen-based habits that your family uses to meet that goal in an average week. Are you surprised by any of your regular habits? Use the third column to check off or write in any alternative habits you already use, or want to add to your family screen time plan. This tool can be used for personal self-reflection, or a larger conversation with your family around shared screen time values.

Screen Time Swaps

SCREEN ALTERNATIVES FOR 0-2 YEARS

PARENTING GOAL	SCREEN HABIT (IN AN AVERAGE WEEK)	SCREEN-FREE ALTERNATIVE
EDUCATION	<ul><li>High quality educational TV show</li><li>App for children</li><li>Electronic educational toy</li><li>------------------</li><li>------------------</li></ul>	<ul><li>Book, game, or song with adult</li><li>Tummy time</li><li>Sensory exploration</li><li>Windows and outdoor walks</li><li>Supervised free play</li><li>------------------</li><li>------------------</li></ul>
FIDGET TOY	<ul><li>TV remote or other handheld technology</li><li>Phone (turned off or locked)</li></ul>	<ul><li>Toy remote/phone</li><li>Teething toy</li></ul>
CO-PLAY WITH CAREGIVER	<ul><li>Educational apps</li><li>Made-for-baby videos</li><li>Selfies on phone</li><li>------------------</li><li>------------------</li></ul>	<ul><li>Interactive songs and games</li><li>Music through a speaker (no screen component)</li><li>Mirrors</li><li>------------------</li><li>------------------</li></ul>

CHAPTER 6
AGES TWO-TO-FOUR: FIRST INTRODUCTIONS TO THE WORLD OF TECHNOLOGY

The toddler and early preschool years are such an amazing time of growth for your little one. It may feel like in the blink of an eye you've traded bottles and swaddles for childproof locks and play dates! As children become more interactive and attentive every day, you may be excited to introduce short screen and technology experiences to spark their curiosity. They're finally at the recommended age when they can truly engage with shows and games! But now that the world of technology is officially open to your toddler, how do you select content and choose what you actually want to share? Are "educational" shows actually better for toddlers? Will your child still want to read books and play outside? How will your impressionable little learner deal with commercials and fast-paced plotlines? In this chapter, we explore recommendations for children's earliest exposure to playful and educational technology use.

Avoiding the Digital Babysitter

It may sound funny, but just like everything else in their world, children actually need to be taught how to watch TV. Research shows that they benefit from being shown and taught how to engage with technology in a healthy way. Enter "active parental co-viewing": the gold standard for screen exposure in early childhood. Research shows that co-viewing content with an attentive, engaged adult helps children understand audiovisual information as an organized story sequence and supports their interactive engagement. Children don't instinctively know that they can talk back, wave, etc. in response to screen-character invitations. With parent encouragement, this kind of physical and social play dramatically increases the educational benefits of screen viewing. In addition to mitigating harms from passive screen engagement, co-viewing has even been shown to support a secure parent-child attachment relationship, simply by reinforcing engaging and collaborative family play experiences. In fact, the APA and other pediatric health organizations recommend exclusive co-viewing of any screen content for children under age five.

You may be thinking, we've just shared some good news and some bad news. The good news is, screens don't have to be as harmful for your toddler or preschooler as you may have thought! Co-viewing can resolve a host of challenges with screen time, simply by providing your safe and secure presence for children alongside their technology-mediated adventures. Of course, the bad news is, experts still don't advise using the screen for one of its most common use cases—a digital babysitter. We recognize that for many families, these guidelines may feel challenging or even unattainable. Especially in the post-pandemic parenting world, what's a mom to do if her family has already developed a passive screen habit?

In the next sections, we describe what the research recommends for keeping your screen time diet as nutritious as possible, and we also share our mom-approved tips and tricks for limiting the screen time battle.

Baby steps: First shows and stories

TV is often the first tech exposure for young children, which is understandable since it's more social and whole-family inclusive than phones or computers. TV can feel like a slippery slope, but if you keep viewing sessions brief and balance their time with other activities, it can be a fun and relaxing experience for your two to four-year-old. Stick to a few trusted habits when selecting screen media.

First, if you are streaming shows, try to use services that limit commercials or let you apply child-mode preferences. At 30-60 seconds, ads are often just the right length to grab children's short-spanned attention, but are often targeted at much older audiences, with bright graphics and adult-themed scenes. Children at this age are especially impressionable, and should be protected as much as possible from adult-facing marketing. Second, it's always a good idea to watch shows together, or view them on your own first before sharing them with your little one. Choose content your child actually enjoys and wants to sing, dance, or play along with, and try to find stories that are appropriate for young children, with real-life actors or puppets, and simple plots, sounds, and images instead of overstimulating animations. Cartoons aren't necessarily worse than live-action, but research on cartoons reveals that certain animations and their corresponding voice actors can unintentionally reinforce negative stereotypes (e.g., foreign accents on classic villains) and misrepresent different ethnic and gender identities to impressionable audiences.

For a first technology experience, you might consider an alternative technology experience to a traditional TV show. Many streaming services now offer "slow TV" options, such as recordings of fireplaces, nature cams in public parks, or scenic landscape train journeys. Other options include gentle wildlife shows or animal-cam live-feeds, which show recordings of exciting animals in natural habitats. Explore.org offers dozens of live cams at nature sites (many of them part of ecology research projects) with "highlight reels" of the best footage of wild animals set to soothing instrumental music. To an adult, these video shows can feel like glorified screensavers, but you might be surprised how it holds the attention of your young child! Most kids are naturally curious about animals, and the activity in natural environments is typically better paced for them to actually follow along with the action. Another idea is to share simple home-movies and candid clips of your own family members. All three of our kids absolutely love watching videos of the family pet or vacation clips from the last trip to visit the grandparents, and it's a great way to reinforce social bonds in the real world through screen time. One note is that toddlers can tell the difference between photos and videos, and the realistic action in videos is typically more exciting to them. We got into the habit of taking one or two ten-second video clips whenever we take family photos together for quick screen time entertainment later on.

How Does my Toddler Already Love TV?

At 15 months, my daughter had already started playing with the TV remote, and pointing and signing "more" at the TV in our living room. We rarely let her even stay in the room when the TV is on, but kids somehow have radar for those super-engaging screens! It seems like after one splashy toy commercial when the TV happened to be on, she was hooked.

Fortunately, she was still young enough that we could redirect her attention to off-screen activities, but occasionally we turned on the TV to extremely low-stimulation shows for about 10 minutes while she sat on our lap on the couch. Her longtime favorite was Dory's Reef Cam on Disney+, a gentle cycle of underwater scenes with no talking and simple water sounds, bubbles, and movements from friendly cartoon fish from the movie Finding Nemo. It honestly wasn't so different from the fish tank at her pediatrician's office.

– Amanda Strawhacker

If your child is ready for a plot-based story, classic shows like *Sesame Street*, *Mr. Rogers Neighborhood*, and *Reading Rainbow* are well-loved for a reason, sharing age-appropriate content through live-action media while meeting the needs of short attention spans. Research also confirms that this type of content is the most likely to be educational and easy for children to understand. Another great idea is to pick shows that were originally books, like *Arthur*, *Winnie-the-Pooh*, and *Llama Llama*. Usually, these plot lines are already interesting and well-written, and it gives you an easy connection to an off-screen activity with a physical book. Many of these shows also have thoughtfully designed apps and websites to extend the fun. Finally, try to co-view shows as much as possible to keep the experience social and interactive. *Bluey* is a great example of a short show (~seven minutes per episode) with as much content for parents (modeling how to resolve sibling differences, immersing yourself in children's play) as storyline for the kids. Revisiting favorites from your own childhood can be another fun way to match your child's interest level in a shared screen experience!

Toddlers, Touchscreens, Apps, and Games

When it comes to interactive apps and games, your two to three-year-old will be a bit limited here. Firstly, you'll want to stick with touchscreens instead of computers. Unlike a mouse, which requires a rather sophisticated understanding that your hand moving on the table matches the movements of the mouse cursor on the screen, a touchscreen relies on much more intuitive gestures and touch pressures. However, what a tablet *can't* offer is the real-life textures and haptic (that means touch-based or vibration) feedback that you can feel when you hold something real. When you squeeze playdoh, you can feel it squish in a satisfying and very real way. Because younger toddlers are still building their ability to predict how something will *feel* based on how it *looks*, common digital feedback like vibrations and blinking lights will still be a bit confusing to your child. You're probably looking mostly at simple tap and drag games on a tablet, such as any app that lets children press simple colors, shapes, and images. Popping virtual balloons, "fingerpainting" on an art app, or playing an on-screen piano would all hold children's attention for a short time.

Older three-to-four-year-olds will probably be ready for more content-themed apps that let them explore preschool topics such as weather and nature (Toca Nature, MarcoPolo Weather); shapes, letters, and numbers (Metamorphabet, Busy Shapes); myself and community (Daniel Tiger's Grr-ific Feelings, My PlayHome); and STEM and non-fiction reference videos made for young children (MarcoPolo World School, Peep Ciencias Epic Videos). Older children are also ready to use the fine motor and spatial skills required to use a mouse and keyboard. When choosing apps for your preschool-aged child, remember that it's okay to offer some apps purely for entertainment. But if you are hoping for an educational experience, ask yourself if the digital experience is actually different or better than an off-screen experience. Some apps are little more than digital flashcards for kids. If you're not sure where to start, local libraries often have computer labs in the children's section preloaded with specially designed products and app suites that are educational for a variety of ages. In addition to testing out apps and software that might be inaccessible at home due to steep school or institutional paywalls, library computer time is a fun and low-stakes way to learn what kind of digital experiences your child prefers, so that you can make better selections on your home devices later on. Finally, remember that all children can still benefit from co-viewing at any age, and the AAP recommends it exclusively through age five years. That guideline may not work for all families all of the time, but it's a good reminder that if you are able to play along with your child, you can support their play while bonding over a new way to interact with the world of technology.

Finally, it's worth looking into the accessibility features that your touchscreen offers to see if you can avoid any issues before they arise when you let your child "drive" the tablet. For example, most touchscreens have an accessibility mode feature that lets you control which parts of the screen children can interact with while they use certain apps. On iOS devices, this is called Guided Access, and Android has a similar feature called Restricted User Access. With Guided Access turned on, you can select any portion of the screen to lock for the user while they use that app, until you turn it off with your passcode. The benefit of this feature with young children is that they can't navigate away from the app you choose, or accidentally close and exit the app they are playing with. In the table below, you can find a list of accessibility features that are great for young children, both typically developing and with special needs.

Best Touch Screen
ACCESSIBILITY FEATURES FOR YOUNG CHILDREN

IOS FEATURE	ANDROID FEATURE	FEATURE LETS YOU...	USEFUL FOR...
Guided Access (iOS)	Restricted User Profile	Temporarily limit access on the device to only one app and only certain areas of the screen at a time	Focusing children on your chosen app without navigating away
VoiceOver	TalkBack	Narrate your child's commands on the device	Auditory feedback to child about what they are doing
Speak Selection	-	Change the rate of speech and highlight words.	Prereaders, second-language learners, and children with language impairments
Dictation	-	Translates speech to text.	Prereaders, second-language learners, and children with language impairments
-	Captions	Offers closed captioning in different modes (speech, text, and style)	Prereaders, second-language learners, and children with language impairments
Invert colors	Color inversion	Reverse all colors in all apps on the screen	Improved screen clarity with higher contrast, and overall less bright light directed at child's eyes
Zoom	Magnification Gestures	Enlarge anything on the screen.	Preventing eye strain
iOS and Android devices all connect to hearing aids and Braille devices via Bluetooth			Families with hearing impairments

Source: www.commonsensemedia.org

A World of Tech Beyond the Screen

Your toddler is hardwired to notice, explore, push, prod, play, and generally get into everything they can. It should be no surprise that they quickly become super curious about screens and technology. So how can we honor that curiosity, while still limiting their exposure to the recommended small doses with adult co-viewing? As adults we're used to thinking of screen time as something that has to happen with, well, a screen. But children love to use pretend play, social play, and stories to help them understand experiences in their everyday life. We can tap into all their natural play instincts to help them understand screens—and we can do it without any added screen time!

TV Basket

If you find yourself surprised at how quickly your child becomes attracted to a specific character or cartoon (possibly before they've ever even seen the show), you are not alone! The marketing geniuses who run children's TV and movie franchises make sure that children don't actually need to come into contact with screens to fall in love with characters. Try this experiment: Walk around your house, nursery, or playroom and try to identify all the franchise characters you can find. Are your child's diapers and clothes patterned with Sesame Street friends? Do you have Disney princess Valentine's Day cards displayed on the fridge? Marvel superheroes on your bags of Halloween candy? You might find it funny or even disturbing how deeply these commercial characters can embed themselves in our everyday lives. But, as every mom of a toddler knows: if you can't beat 'em, join 'em.

Leaning into your kid's love of Elmo doesn't mean you need to integrate more video time into your day than you feel ready for. Instead, why not try a "TV basket"? You may have seen this trend online (because it really does work). A TV basket is a small basket of books, toys, and off-screen activities that are branded with your child's favorite TV character. Even if you never watch Sesame Street on TV, Elmo can be just as exciting in a picture book, as a stuffed animal, or singing a song from a screen-free toy. Recognizing and naming these characters while you're out shopping or on a walk can become a fun game for your child to build bridges of familiarity across many contexts of their day. When they're old enough to watch TV, it will likely make the viewing experience richer if they can connect the on-screen stories with their long-time real-life playmate!

If you feel uncomfortable with the idea of fostering a love of commercial screen-based characters, there are hundreds of non-screen fictional characters to love as well. Characters from local storybooks are always fun, since they are more likely to be depicted in neighborhood shops, parks, and playgrounds where your child can spot them. For example, the city of Boston has so thoroughly embraced Robert McCloskey's classic picture book, *Make Way for Ducklings*, that you can find duckling artwork in local book shops and libraries, wall murals at the Boston Logan International Airport, and there is even a commemorative statue of the ducklings in the city's major downtown park, the Public Gardens. Spotting familiar characters in the "wild" is a fun way to help children see reflections of their personal, emotional experience in the larger world around them, and to help them feel a sense of belonging as members of their wider community.

Audio Tech Time

Along the same screen-free lines as a TV basket, you might find that audio technology becomes your child's new favorite activity. There are hundreds of podcasts, soundtracks, and radio shows out that are specifically designed for very young listeners. You might try getting some podcast story time in during a long car ride, playing their favorite music playlists while they're playing with play-dough or paints, or even play a favorite TV show on your phone with the screen facing away from your child while they engage with their TV basket. As with all technology, co-viewing (or in this case, "co-listening") to audiobooks and podcasts is a great way to build a shared vocabulary of favorite stories, fictional friends, and play experiences.

To infuse audio tech time with more autonomy, older toddlers will enjoy audio players like Tonie Box or Yoto Player. Marketed for ages three-plus, Tonie Box is a small speaker that uses figurine toys encoded with small electronic chips to activate the box and play different bedtime stories, songs, and character voice recordings. It uses simple actions to change the volume and the audio tracks, completely foregoing the need for a screen. This is a great way to extend off-screen play if your child has specific movie or book characters they love. It can also bridge the gap from reading with an adult to reading independently, with pre-readers playing their audiobook and following along with the same picture book. Part of the beauty of audio players is that kids can pick their own story experience by selecting a toy to "play" with, giving toddlers that feeling of being "in control" that they so desperately seek. These story players are designed to "age up" with your toddlers and can be used well into elementary school by choosing longer and more complex stories or chapter books to listen to.

Pretend Play with Screens

Sometimes when it seems like a toddler really wants to play with a screen, you can satisfy their curiosity by offering some technology-themed pretend play instead. Here are some things that our kids accepted as stand-ins for screen time in the toddler years:

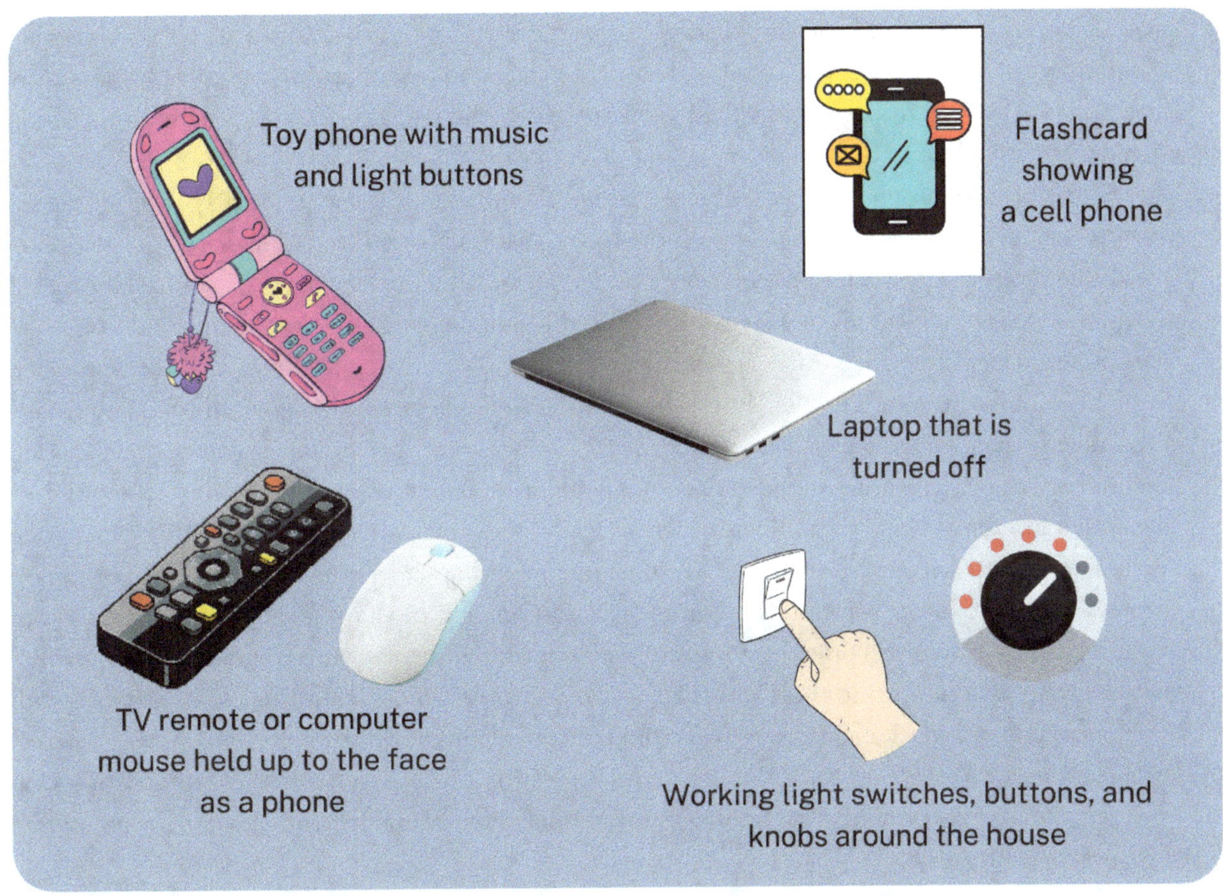

We aren't proposing that you try to trick or misdirect your child about screens. In fact, most children under age two can tell the difference when a device is powered on or off. However, in our experience, most of the times when a young child is seeking some kind of technology experience, they are satisfied with a *pretend play* version of that experience. Just like pretend cars, kitchens, workbenches, and other common imaginative play scenarios that children see at daycares and museums, pretend technology spaces offer children the chance to practice doing the things they see adults in their lives doing. Typing on a computer, talking on the phone, scanning barcodes at a cash register, using electric keys or remotes, and even playing videogames are all activities that children typically observe their family and friends engaged in on a weekly or even daily basis. Any attempt to try these things out on their own is part of their natural drive to learn about social behaviors through imitation and play.

Sometimes, as moms, we can feel concerned or annoyed by this type of play when it comes to screens. Technology isn't the only kind of play that elicits this reaction from adults. Seeing children engage in exploration of their own body parts, trying out curse words, and asking about drugs or alcohol can also make us feel uncomfortable as caregivers. But just as it wouldn't be safe for children to overindulge their curiosity about technology through unlimited screen-time, it's arguably not healthy to completely restrict and shut down their natural curiosity about this common feature of their physical environment. Encouraging safe pretend play options around technology use can be a harmless way to encourage their exploratory drive, even to the point of letting children play with technologies themselves as toys in a safe and supervised way. Anyone with a toddler obsessed with cars knows that they could spend hours sitting in a driveway on mom's lap in the driver's seat, turning the wheel and jingling the keys. You might find that your little tech guru wants to "drive" a computer the same way, or play with a videogame controller while sitting in your lap and the TV turned to a screensaver. Kids love to become obsessed with things, and then just as quickly, move on from them to a new shiny obsession. As long as you feel comfortable with the level of exploration, and your little one is safe from any real hazards, you can rest assured knowing that technology pretend-play is a natural byproduct of growing up in a technology-rich world.

Pretend Play with Screens

When my daughter was a year and half, we got her first library card and started taking regular trips to the library to play and read together. She always loved the library, but once she noticed us checking books out with the library computer scanner, she became so curious about the computers there.

In our children's section, there is a desk with several desktop computers and color-coded "learn to type" keyboards. The computers are all pre-loaded with PreK-appropriate learning software. There is also a lower tabletop touch display for collaborative games with groups of children. My toddler was obsessed with these devices, and it seemed like every time we visited the library, all she wanted to do was sit and tap on the screens. I didn't mind at first, but soon I was getting tired of the meltdowns when I asked her to leave, and wondered if it was even worth coming to the library at all.

Dipping a Toe in Tech

We hope you found the tips in this chapter useful for helping your two-to-four-year-old begin to experience the world of technology in a safe, healthy, and respectful way. All the same rules of toddlerhood apply, and tantrums are an opportunity to set whatever firm boundaries you need to set around technology use and screen time (see Chapter 11 for more tips about that). But hopefully you also saw some easy ways to allow children to explore their natural curiosity around technology, and even take control of some early experiences with tech play, whether it's screen-based, off-screen, or in between!

TEACH YOUR TODDLER HOW TO WATCH TV THOUGHTFULLY

BEFORE WATCHING

Decide on the type of show

- Home movies: Video clips from family vacations, clips of daily activities, clips of pets, etc.

- Slow TV: Nature cam, still recording (e.g., fireplace, aquarium), railway journey.

- Story-based shows: Opt for live-action, book adaptations, or classic educational favorites from non-profit creators.

Review your selected show ahead of time

- Check a trusted ratings source, like Common Sense Media.

- Watch the show yourself.

WHILE WATCHING

Co-view with your child whenever possible

Model how to engage actively as a viewer

- React to the show when it invites play invitations, greetings, gestures, sing-alongs, etc.

- Ask your child questions and make comments about what you see on screen.

- Take chances to break eye contact with the screen to make eye contact with your child. Playful talking and physical contact (like hugging) can support this.

Limit sessions to keep viewing time short and sweet, and avoid overstimulation

- You might use a phrase ("night night, TV!") or a short five to ten second song to mark the end of TV time, and always turn off the TV when done.

- Keep the ending of TV time light and upbeat, to avoid children associating this transition with negative feelings.

AFTER WATCHING

Immediately: Plan a fun offscreen activity that's ready to go right away to redirect attention

- If kids want more time with the show, switch to a TV basket to keep the play going off-screen using themed books, stickers, dolls, etc.
- Some kids do better with a total state change: move to a different room, head outside for a walk, or take a bath.

Later on: Recall favorite moments from the show that you viewed together to model mindful viewing

- Talk about what you watched together as you reflect on your day at bedtime.
- Read or take out books from the library related to topics from the show.
- Sing songs or retell jokes from the show during regular play activities.

CHAPTER 7

AGES FOUR-TO-SEVEN: RAISING CREATORS (NOT CONSUMERS) OF TECHNOLOGY & MEDIA

Let's play a quick game of "one of these things is not like the others" to set the stage for our chapter. Which of the following does NOT belong in this group:

a) Computer
b) Television
c) Tablet
d) Paintbrush

What did you select? If you chose d) paintbrush, then you answered how most people would. A paintbrush seems obvious: the other items in this list are all digital technologies, while a paintbrush is not. Simple.

That makes sense, yes, but there is another way to answer this question. We would choose b) television when answering this question. Okay, let us explain. Options a, c, and d are all tools that allow users to create, design, and express themselves. They all allow users to engage *actively*, as *creators* of media (whether it is print media with the paintbrush or digital media with the tablet and computer). Except the television. For the most part, television engages users only passively, not actively. Television allows children to *consume* media but not *create* their own.

Children in pre-kindergarten and early elementary school are practicing amazing skills every day, many of which are related to actively creating and communicating. They are learning to read, write, draw, design, speak, and organize their thoughts and ideas in effective ways. They are using a variety of tools, including paper, pens, books, and open-ended supplies, to help aid in their communication and self-expression. Why, then, should their use of technology be any different? This chapter will explore all the ways that young children, around ages four to seven, can begin harnessing digital technology to be creators rather than consumers of their digital experience. We will talk about coding, computer science, and computational thinking as a new form of literacy and share ways that, beginning in preschool, parents can support kids' foundational coding skills. We will also look at how we can integrate the arts with technology to help young children harness technology as a means of creativity, communication, and self-expression.

Empowering Digital Creators

Information and Communications Technology (ICT) or digital technology has been a part of educational systems in the U.S. and around the world for a long time. It has often been used to help support students' mastery of new concepts, drill and practice skills (think letter recognition or basic math facts), or as a tool to research and discover new information about a variety of subjects. But as technology has grown in prevalence in schools and at home, especially with young children, some researchers and educators have also expressed concern that excessive usage of computers and digital technologies may actually stifle children's learning and creativity through passive consumption of media. In order to address these concerns, there is a growing body of work from notable researchers such as Mitchell Resnick (MIT) and our own mentor Dr. Marina Umaschi Bers (The DevTech Research Group) on how technology can be used to foster positive behaviors and engage children as *creators* rather than *consumers* of digital content. One way to do this is by explicitly integrating the arts, self-expression, and identity exploration with traditional technology and engineering curricula for young children.

What does it mean to be a digital creator for children ages four to seven? Well, in most places in the United States, children in this age range are in pre-kindergarten or early elementary school, and therefore just embarking (or about to embark) on their formal education journey. ICT and technology education have often fallen under the Science, Technology, Engineering, and Mathematics (or "STEM") umbrella when thinking about formal education content. Adding the arts to STEM-based subjects, such as computer programming and engineering, may enhance student learning by infusing opportunities for creativity and innovation. The acronym "STEAM" (Science, Technology, Engineering, *Arts*, Mathematics) expresses the concept of promoting creativity and expression through technology and the sciences. The "A" for art in STEAM can represent more than just the visual arts, but rather a broad spectrum of the humanities including the liberal arts, language arts, social studies, music, culture, civics, and more. The takeaway is that art, of any kind, can and should be explored alongside topics like engineering and technology. By doing this, we help to encourage creative ways of using technology, over passive consumption.

Wondering how you can support your young children at home so that they are engaging with technology as creators rather than just consumers of media? Here are a few simple activities that infuse the arts with technology that you can try with your own young children:

- **Photography:** Taking and editing digital photos
- **Filmmaking**: Recording and editing short movies or animations
- **Digital Art:** Creating digital artwork or collages
- **Digital Presentations:** Creating a digital presentation to share facts or ideas on a subject they are learning about

Computer Science, Coding, and Computational Thinking: What Does It All Mean?

When it comes to supporting our children as digital creators and supporting the skills that will propel them through school and their careers, coding is probably one of the most critical things to learn. This can be daunting to parents and caregivers who did not grow up learning to code themselves. So where do you even begin? You may have heard about initiatives like Hour of Code, Computer Science Education Week, Girls Who Code, and more "big name" ventures over the last decade but you might be wondering what all the fuss over learning to code is about. Well, in our own research, we have seen many benefits to teaching coding beginning at a young age. For example, we have found that learning to code helps to support young children's sequencing skills (an important pre-math and pre-literacy skill), problem-solving skills, can prompt collaboration and other social skills, and more. Research has also shown that children who are exposed to STEM curriculum and programming, such as computer science, at an early age demonstrate fewer gender-based stereotypes regarding STEM careers and fewer obstacles entering these fields later in life.

Some of the many skills kids gain when learning to code are computational thinking skills. Computational thinking (CT) is a creative way of thinking that enables individuals to solve problems in systematic ways. While rooted in the field of Computer Science, CT is more than just relevant to technical fields; it is a skill set that encourages us to solve problems, develop effective solutions, and think logically–and creatively–as we encounter everyday dilemmas. CT is increasingly being recognized by researchers and educators as being just as important as reading, writing, and mathematics. Before you worry that this is something you were not aware of, if you have a young child at home, you are likely practicing CT with your children already (whether or not you realize it). This CT practice is happening through retelling familiar stories using sequential language, encouraging them to follow step-by-step instructions, and helping them break down big tasks into more manageable steps. Simply by being intentional with the language you use, the questions you ask your kids, and the activities you facilitate can deepen the way you are already supporting your children's CT learning and growth.

Key Coding Definitions

Having trouble keeping all these new high-tech terms and skills straight? Here are a few easy definitions to help you keep track of it all.

Computer Science - Computer science is the study of computers and computing as well as their theoretical and practical applications. It may build on mathematics and engineering skills and principles.

Coding - Coding is the act of creating a list of instructions for your computer (or robot, tablet, or other technology) to follow. These instructions determine what actions a computer can and cannot take. Coding involves writing code in a language that both humans and machines can understand.

Programming Languages - A programming language is a way for people to communicate with computers and other technologies. There are adult programming languages you may have heard of like JAVA or C++ and also graphical programming languages that can be used by children like Scratch or ScratchJr.

Computational Thinking (CT) - Computational thinking is a creative way of thinking, rooted in Computer Science, that enables individuals to solve problems in systematic ways. It includes skills like debugging, algorithmic thinking, and more.

Supporting Computer Science Education with Pre-K through Early Elementary Kids

Now that you know what coding is and why it is important to introduce it to young children, you may be wondering how best to foster this introduction to your own young kids. Our mentor and founder of the Developmental Technologies Research Group, Dr. Marina Umaschi Bers, describes coding as literacy. Developing literacy involves gaining new skills and ways of thinking about ourselves and the world. The same is true for learning to code. Very young children can explore coding through hands-on (screen-free) tools such as programmable robotics kits (we share a bunch of examples in Chapter 10 of this book, which is all about screen-free technology!), coding-themed board or card games, picture books, and more. But there are also a lot of valuable digital apps and experiences that can support children's learning to code in an open-ended way. One of our favorite examples is ScratchJr.

ScratchJr is an introductory programming language that enables young children (ages five to seven) to create their own interactive stories and games. Children drag and snap together graphical programming blocks to make characters move, jump, dance, and sing. Children can modify characters in the paint editor, add their own voices and sounds, even insert photos of themselves—then use the programming blocks to make their characters come to life.

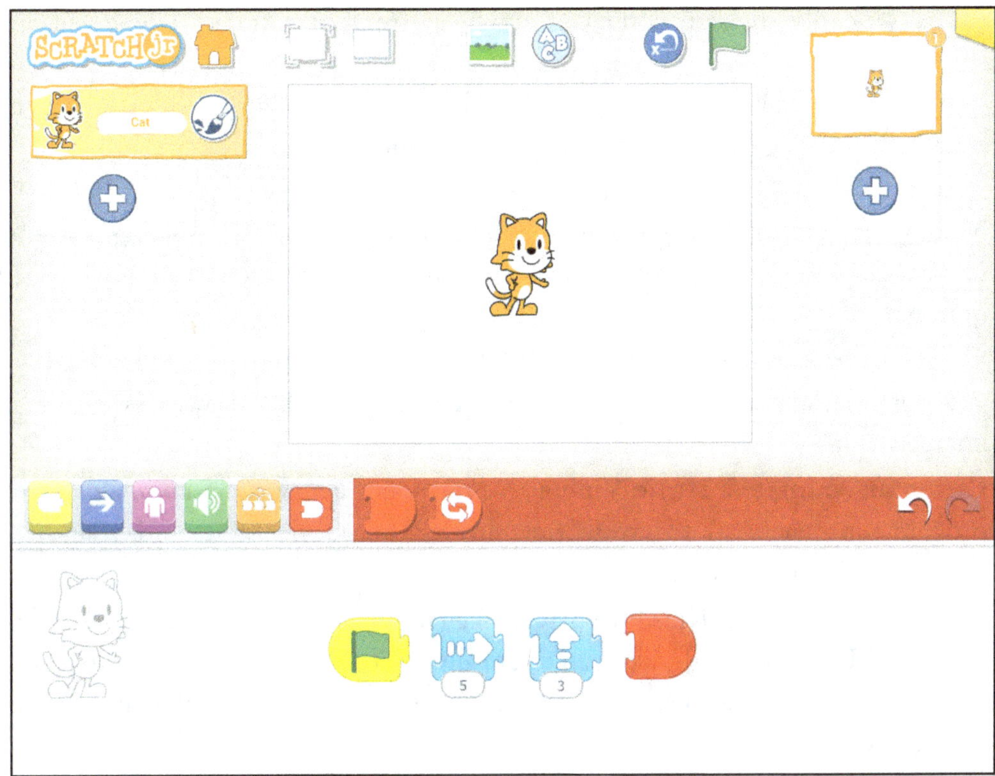

ScratchJr was inspired by the popular Scratch programming language (check out: scratch.mit.edu to learn more), used by millions of young people (ages eight and up and even sometimes in introductory high school and college coding classes) around the world. ScratchJr redesigned the Scratch interface and programming language to make it developmentally appropriate for younger children, carefully designing features to match young children's cognitive, personal, social, and emotional development. But best of all? ScratchJr is completely FREE! It is available for free download on iPads, Android tablets, and Chromebooks (learn more at www.scratchjr.org).

While many people envision coding as sitting in front of a screen typing out strings of numbers, it can be a very creative process, especially with young children. Here are a few of our favorite ways to use apps like ScratchJr in a STEAM context that integrates the arts in different ways that embody creating (rather than consuming) technology:

- **Mixed Up Fairy Tales:** Integrate reading and language arts by creating new endings to familiar fairy tales (or any stories you are reading!), using ScratchJr's multiple-page project option.

- **Create an Interactive "All About Me" Collage**: Get creative using ScratchJr's camera and paint editor functions to code an interactive project that showcases what makes you uniquely you.

- **Create an Animated Music Video:** Use the sound recording feature along with other programming elements to create an animated music video! Record yourself singing, playing music, and more!

- **Create a Videogame:** Kids love playing digital games on their tablets, so why not create your own? Using ScratchJr, kids can create all sorts of interactive games using some pretty advanced coding concepts in the process! This makes for the ultimate digital creator experience.

If you are still feeling nervous because you are new to coding, remember that you can learn alongside your children using the same digital tools and apps as they are. Show them that it is okay to make mistakes and role model the fact that you are a lifelong learner. You might even surprise yourself and find yourself interested in learning more!

Creators vs. Consumers Checklist

Remember, not every tool or app you put in front of your child needs to be a "creator" tool. However, it is important for adults to be aware of how many educational apps and technologies their children use that offer only passive engagement. To raise well-rounded digital citizens who are not only able to *use* technology but also able to *create* with and *understand* technology, we must foster the right type of play and learning from an early age. Here is a simple checklist of digital features that are common in creator and consumer technologies.

Identifying "CREATOR" VS. "CONSUMER" TECH EXPERIENCES

TECHNOLOGY FEATURE	CREATOR	CONSUMER
Allows for open-ended exploration	✓	
Allows for creating your own media (pictures, animations, movies, music, etc.)	✓	
Engages you in open-ended coding experiences	✓	
Allows for open-ended art creation	✓	
"Sandbox" style games that allow for creativity and exploration without a predetermined goal		✓
Allows you to watch but not create media (videos, animations, movies, etc.)		✓
Prompts you to progress through very structured "levels" or modes of play		✓
"Drill & Skill" apps or games that prompt you to practice the same skills repeatedly to improve their accuracy		✓

Now, take some time to explore your children's favorite digital games using the Self-Reflection log provided below. Do they have more creator features or consumer features? How can you find more balance in the types of digital activities your child explores? We will explore this further in Chapter 9 when we discuss meaningful screen time exposure.

Children's Apps & Games
REFELCTION LOG

APP/GAME NAME	CREATOR OR CONSUMER TECH?	PERSONAL NOTES

PART THREE

Becoming a Tech-Savvy Mama

CHAPTER 8

TECH-SAVVY MAMAS & BREAKING GENDER STEREOTYPES

Let's kick off this chapter with a quick little game of "Two Truths and a Lie". Read the three statements below and try to guess which one you think is a lie:

1. Moms, regardless of educational or career background, are pivotal to their children's success with technology and STEM in general.
2. Moms, regardless of educational or career background, can counter pervasive gender stereotypes related to technology and engineering.
3. Moms need a solid background in technology or engineering in order to counter gender stereotypes related to technology and engineering.

We hope you chose statement number three as the lie! Yes, if you are a woman with a technology or engineering-related career, you can be a powerful role model, breaking stereotypes for your children. But guess what? ALL moms have the power to serve as powerful role models when it comes to technology, engineering, and STEM/STEAM more broadly.

This chapter will get into all the specific ways that moms can effectively model interest and capacity for technology, coding, engineering, and more. It will also introduce other strategies for showcasing diversity in the fields of technology, computer science, and engineering through the books, media, and activities you share with your young children at home. But before getting into those specific strategies, we will discuss women's representation in technical STEM fields and why this is an issue moms should be aware of, to begin with. We will explore foundational research and information about stereotypes, including when and how they develop, and what moms can do to counter them.

Girls' and Women's Representation in Tech Fields

Women, particularly women of color, have been historically excluded from many STEM fields. Women make up nearly half (around 48%) of the workforce in the U.S. but only 34% of the STEM workforce. Black, Hispanic, and Indigenous women represent less than 10% of the STEM workforce. The underrepresentation of women is starkest in certain STEM fields such as computer science, engineering, and physical science where women make up only around 26% of computer and mathematical sciences and 16% of engineers.

So what causes this persistent underrepresentation? We know from extensive research that this underrepresentation is **NOT** because of differences in skills, abilities, or aptitude, but rather barriers that girls and women uniquely face, operating within larger systems of oppression. These barriers include things like stereotypes and biases, a lack of culturally or personally relevant STEM curriculum, a lack of exposure to female role models in STEM, and more. These barriers often discourage and hinder girls' identification with STEM and even influence women's career trajectories and long-term participation in STEM. For women who *do* overcome these barriers and pursue a career in STEM, they end up facing new barriers such as workplace discrimination, inequity in pay and promotions, and inadequate parental leave, just to name a few challenges (maybe some of you moms reading this book are in the thick of this right now!).

We realize this section might feel distressing to you right now, but we promise, there is good news coming! Research has shown there are a lot of ways that we can overcome the barriers facing girls and women in STEM. And guess what? A lot of it starts with YOU as a mom, and the mindsets, beliefs, and approaches you foster at home with your little ones right now. So we could not let a book on technology and motherhood go by without providing you with some practical tips and resources for addressing the barriers described earlier and breaking stereotypes for the next generation.

Exploring A New Model of Motherhood: Combating STEM Stereotypes

If you are reading this book, you are presumably a mom-to-be or the mom to a baby or young child. So you might be thinking you have all the time in the world to consider the development of stereotypes and your child's interest in technical STEM fields. In reality, stereotypes begin to form early—probably earlier than you may think (we'll get back to this in just a second)!

You might also be tempted to skip this chapter if you are a boy-mom. After all, isn't the focus of this chapter role modeling and reducing barriers for girls and women? This, too, just isn't the case. It is incredibly important for all young children—regardless of sex or gender identity—to be exposed to positive messaging around girls and women in STEM. If you are a boy-mom, we are pleading with you not to skip this chapter. We desperately need the next generation of boys to grow up without the same stereotypes and biases as previous generations. We know this can be a heavy burden to carry, but the beliefs and mindsets that our children hold about women start with mom.

Okay, back to stereotypes. A "stereotype" is a widely held but fixed and oversimplified image or idea about a specific group. Stereotypes may relate to gender, sex, race, ethnicity, age, professions, hobbies, and more. Whether or not you believe a stereotype to be true, it can still have a profound impact on your performance and identification with STEM.

Beginning at an early age, stereotypes have the power to influence children's interest, confidence, and identification with STEM fields. Researchers from New York University, for example, found that by age six, girls are already likely to believe boys are smarter. When it comes to technology-related stereotypes, recent research by Dr. Allison Master, Dr. Andrew Meltzoff, and Dr. Sapna Cheryan, found that stereotypes about girls' interest in computer science and engineering are formed as early as age 6 and are evident across multiple ages from childhood through adolescence.

The good news is there are lots of ways that moms can counter these stereotypes and ensure that all young children are equally encouraged to succeed in fields like computer science and engineering. In the following section, we will explore the role of parents and caregivers in countering stereotypes and encouraging academic success in technical STEM fields.

Stereotypes About Technical STEM Fields

What do you think of when you picture a computer scientist? How about an engineer? A website designer? An app creator? Close your eyes and let yourself imagine the people in these careers without censoring your thoughts.

If you are like many people, you may immediately picture a man. You may also have the stereotyped belief that they are "nerdy," antisocial, or bad at things like teamwork and communication. (You can thank some of the popular TV shows and movies over the past decade for this male, nerdy depiction of the sciences!)

These stereotypes can impact the likelihood that girls and other historically excluded groups pursue STEM fields - even if they don't believe that they are true.

When introducing your own young children to engineering projects, science experiments, and coding, think about how you can broaden some of these pervasive stereotypes. You can showcase STEM as collaborative, social, creative, and FUN! All of this can start at home through early play experiences and the messages that are portrayed by parents and caregivers.

The Role of Parents and Caregivers

It may seem intuitive that parents and caregivers are influential when it comes to kids' confidence and desire to persevere in different academic or extracurricular environments. Well, technology is no exception. Parental support, expectations, and pressure have all been identified in research as key factors for students to pursue a STEM career and to maintain interest in STEM. In research conducted by Google on the factors that influence young women's decision to pursue computer science, encouragement by parents and exposure to out-of-school computer science activities arose as key factors in women's persistence in computer science. Interestingly, some research points to the powerful impact of other women in the family beyond mom, especially older female role models. "Tech grandmothers" who engage with technology can challenge both gender and ageism stereotypes, showing children that curiosity, learning, and innovation are not limited by age or gender.

On top of the *explicit* encouragement and support, it is important for parents to remember that kids are always watching and listening. When it comes to stereotypes and ingrained beliefs about gender roles and abilities, the *implicit* messaging we share is just as important (maybe even more so) than the explicit messages we share. How we talk about our own abilities, interests, and skills (even in passing) is just as important as the explicit messaging we teach our children.

Let's take a moment for self-reflection. Think about how you do (or don't do) the following:

·Do you model your own curiosity about trying out new technology?

·Do you show enjoyment in learning new technical skills?

·Do you play with screen-based technology with your child?

·Do you play with screen-free technology with your child?

·Do you model problem-solving skills when you encounter an unexpected technical problem?

·Do you model healthy boundaries and limits with your own screen-based technologies like your phone?

If you answered "yes" to most of the questions above, you are probably already doing a great job of modeling your competency and confidence with technology and your own comfort troubleshooting technical issues you cannot figure out. If you answered "no" or you weren't sure about most of the questions above, this is an exciting place for you to start! Remember that your kids are always watching and listening to you. Moms are in a unique position to break gender stereotypes and serve as female role models. It can feel like a heavy burden at times, but moms typically play a very large role in how young children formulate their beliefs about women's interests, skills, and abilities. Remember, you don't have to know all the answers or be a coding wizard to help break these persistent stereotypes! Simply showcasing your interest and curiosity about trying (and your comfort in making mistakes!), having fun co-playing with technology, and narrating your problem-solving approaches will go a long way in making a positive impact on your children.

Did You Know?

Personal views about intelligence and failure can impact girls' achievement and long-term persistence in rigorous STEM fields like computer science and engineering. Encouraging a "growth mindset" or the belief that intelligence is not fixed, but instead can change and grow incrementally through practice and hard work, can help counter the stereotype that some people are just "naturally" gifted at math, science, coding, and more.

So how do you talk about the growth mindset with young children? You can simplify this concept by telling kids that the brain is like a muscle that can get stronger with more hard work and practice. You can also normalize making mistakes as part of learning any new skill and celebrate the learning process instead of just the final **product**.

Want to learn more about the power of the growth mindset? Check out the book "Mindset: The New Psychology for Success" by Carol Dweck.

Strategies to Implement at Home

There are other ways that parents and caregivers can help counter stereotypes. Here are some easy-to-implement strategies for breaking stereotypes related to technology and engineering fields and supporting equity at home with your young children:

- **Equally encourage coding and tech toys**—STEM toys, including coding and engineering toys, have a long history of being marketed to boys. Research conducted by the LEGO Group and the Geena Davis Institute on Gender in Media indicates that parents today still more frequently encourage sons to do STEM activities, including using STEM-themed toys, than daughters. This may subtly reinforce stereotypes about who can pursue STEM, and girls may also miss out on important early learning experiences. As parents, try your best to offer high-tech STEM play opportunities equally to all children, regardless of sex or gender. When it comes to coding, robotics, and engineering, ALL young children can benefit from these early play experiences!

- **Encourage hands-on learning**—We have seen the research that stereotypes begin to form early. But research has also shown that children learn best through hands-on play and exploration. It is important to begin reaching all young children with hands-on opportunities to code, engineer, and tinker to pique their curiosity about technical STEM fields from an early age.

- **Encourage mistakes**—Mistakes are a part of the learning process in all fields. This is especially true of STEM fields like engineering and computer science. By fostering a healthy mindset that mistakes are okay and part of the learning process, you are setting your kids up for success!

- **Showcase diverse role models in books and media**—Exposing children to diverse role models can help break stereotypes about who can and should pursue fields like computer science and engineering. Role models can play a powerful role in inspiring children and youth to envision themselves in careers. So what do you do if you don't have any women friends who are computer scientists, engineers, or astronauts that can serve as role models for your children? Look closely at the books and media your children engage with! Are scientists and engineers portrayed in a diverse and welcoming way? Or are stereotypes reinforced? Choose books, TV shows, and movies that showcases the diverse world of STEM and include characters that you want your children relating to and looking up to.

- **Encourage curiosity**—Digital technology, programming languages, artificial intelligence, and robotics will all change year by year. The best way you can support your children in becoming (and remaining) digitally literate is to encourage their curiosity and lifelong love of learning. Support your children in asking their "why" questions and experimenting to figure out the answers. Remember to be mindful about equally encouraging curiosity about technology and engineering amongst all children regardless of gender or background.

Navigating Gender Stereotypes with My Son and Daughter

As a researcher who has devoted her career to investigating and countering STEM stereotypes, I was ready to ensure that I was equally encouraging my son and daughter to explore and engage with technology, coding, and all things STEM from an early age. I have also been extremely mindful about representation of STEM characters in books, TV shows, and other media my kids are exposed to.

Despite all my own best efforts, I began to realize that my young children were still being exposed to subtle stereotypes. For example, my son from an early age has received gifts like LEGO sets, engineering kits, and robots from friends and family members. My daughter consistently receives clothes and dolls from friends and family members. I don't think she has ever received a STEM-themed gift from someone other than my husband or I. They are also exposed to stereotypes anytime we walk down the toy aisle of our local stores, in the greeting cards they receive for their birthdays, and the messaging and conversations they have with their friends outside of our home.

– Amanda Sullivan

Thinking about Screen Time

In this chapter, we examined the many ways that parents and caregivers can counter stereotypes and set their kids up for success in technical STEM fields. One way we discussed was exploring technology and apps that introduce coding and engineering together with your children. Well, this can be a challenge when we are worried about moderating screen-time exposure! In the next chapter, we will explore meaningful screen-time exposure that can be used to facilitate active learning rather than passive consumption of media. Spoiler alert: we'll discover that not all screen time is created equal!

CHAPTER 9
MEANINGFUL SCREEN TIME EXPOSURE

"We have to be role models. We have to be curious. Above all, we have to pay attention
and be part of the conversation while our kids are still listening to us."
– Anja Kamenetz, *The Art of Screen Time*

The time has come! Your child has reached the age and stage where you feel they are ready to begin exploring screen-based technology! But where to begin? And how to decide what works best for you and your family?

How Much? How Often?

The first question that many parents ask about screen time is about the *time* component: how much is right for my child? Fortunately, pediatric health and research organizations can offer a straightforward answer to that. For two-to-five-year-olds, most recommendations suggest no more than an hour a day spent co-viewing educational screen content. At six years and older, experts suggest anywhere from zero to three hours of non-educational screen time. However, the average attention span for a child is around fifteen minutes, and that's even less in younger children. This is a useful starting point as you consider how to incorporate screen time into your child's day.

Screen Time Recommendations
AND AVERAGE ATTENTION SPANS BY AGE

AGE	SCREEN TIME RECOMMENDATIONS	AVERAGE ATTENTION SPAN
2 years old	< 1 hr a day, co-viewing with an adult	4-6 minutes
3 years old	< 1 hr a day, co-viewing with an adult	6-8 minutes
4 years old	< 1 hr a day, co-viewing with an adult	8-12 minutes
5-6 years old	0-3 hours of non-educational screen time	12-18 minutes
7-8 years old	0-3 hours of non-educational screen time	16-24 minutes

Source: https://www.thetreetop.com/statistics/average-human-attention-span

Of course, there's also the individuality factor. Every child is different, and one hour a day can impact one child's sleep, diet, and attention habits very differently from another—even within the same family! We suggest starting small and taking stock. One of the easiest ways to see the effect of screen time on your child's schedule is to break tech use up into small chunks of manageable time. You might start your four-year-old son with one 10-minute chunk of screen time a few times a week, then build to two-to-three chunks per day once he enters kindergarten. By his seventh birthday, you could extend those two-to-three screen time breaks in the day to 20 minutes per session, or find some other way to increase responsibility and time according to his age and maturity.

You'll notice with this example that we're nowhere near that one-to-three-hour guideline we mentioned above. There are two reasons for that:

1) Expert guidelines are safe *upper* limits for how much screen time children can get in a day, but they are not requirements. You can always go lower or skip screens altogether if you wish! And,

2) You may decide that you want to have some extra wiggle room in your daily screen budget. Maybe you want to go on an errand to a boring place where you'd really like to just hand off your phone, or you know that whenever you grab a coffee at your favorite coffee shop, there's a blaring TV that your child loves to sit and stare at. Maybe your child's school uses screen time for 30 minutes a week during math class, or you know your babysitter likes to watch TV episodes with your child on the weekends. By building a bit of cushion into your family's media diet, you don't need to worry so much about all the passive or unscheduled screen time that your child encounters. And believe us, in today's fast-paced world of ubiquitous screens, they *will* definitely encounter screens more often than you think!

How Should Kids Use Their Screen Time?

If there's one lesson we hope every reader takes away from this book, it's this:

All screen time is not created equal!

In general, more active screen use is better for learning and growth than more passive consumption. Watching videos is the most passive, while creating new digital content (movies, stories, coded animations) is the most active, and interactive screen activities like video games and apps usually fall somewhere in the middle, as shown in the figure below. Even within that range, there is nuance. More interactive shows and videos with a slower pace, meaningful messages, and more prosocial content are more beneficial than shows with violence or adult themes. Computer coding is a positive way for children to explore digital creation and expression, but playing a video game on a coding website like Scratch (www.scratch.mit.edu) is not functionally different from playing any other video game, unless kids take the time to click around inside the code and figure out what the commands do. You may be starting to see a trend here: screen time has many forms, but whether screen time is educational depends on what your child is doing while engaging with the screen. That's probably why co-viewing is so beneficial, since parents can observe children's screen viewing habits in real-time, and model how to engage playfully with screens rather than just watching them.

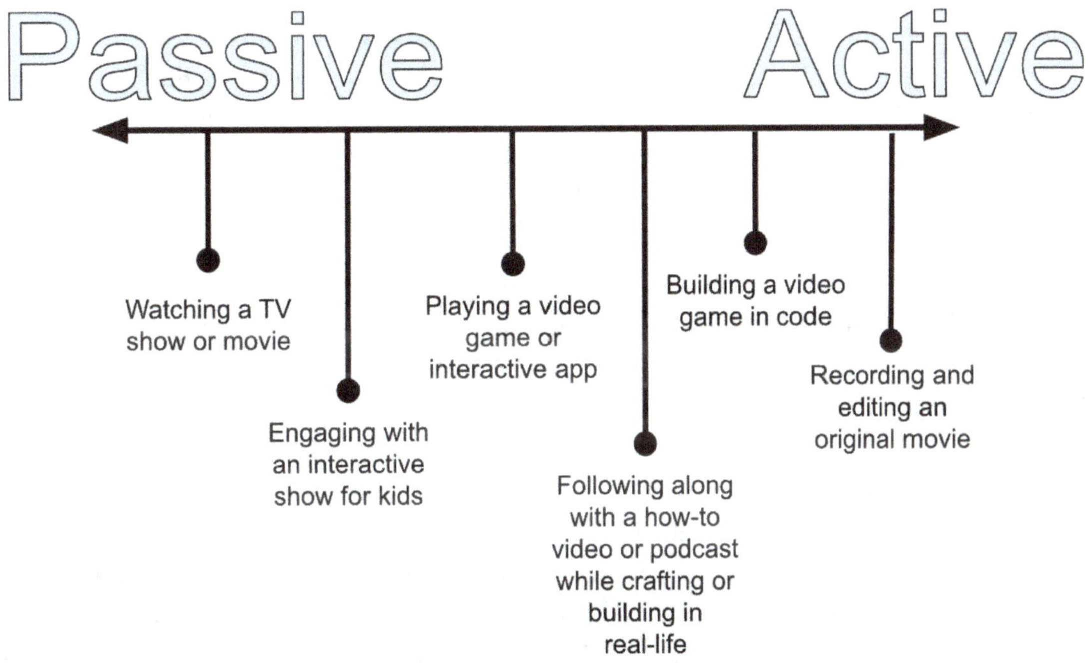

When it comes to selecting specific apps, games, and websites, we always recommend trying things out yourself before letting your child explore. In addition to your own first-hand experience, it can be helpful and time-saving to find trusted reviews and sources to guide you. Common Sense Media is a tried-and-tested review website for parents to help make informed choices about children's digital content, with reviews on everything from children's TV, movies, and streaming shows to podcasts, books, websites, and apps. The experts at Common Sense Media use up-to-date research in their educational resources and parent guides, and their content reviews are detailed, balanced, and fair. Using a reliable source like this can save you hours of scrolling on your phone every time your child comes home with a new favorite app or show.

One of the uncomfortable truths that parents face when selecting content for their children is recognizing that all commercial games, apps, and shows are at their core built for capturing children's attention in order to sell something. Commercials and ads built into content targeted at children can leave lasting and sometimes harmful impressions. Whenever possible, select ad-free versions of your children's favorite content to limit their exposure to these subliminal and repetitive messages. If you're interested in learning more about ways to combat marketing in your child's digital diet, we recommend the Fairplay nonprofit (fairplayforkids.org). This organization is devoted to helping parents understand the unseen impact of commercialization on children's daily lives, learn about the importance of off-screen experiences and activities, and advocate for ways to mitigate commercial branding in your child's play. There are even resources to participate in an annual Screen-Free Week, for a mindful change in your screen routine!

Screen Time for Them = Free Time for Me?

What should *you* do while they're having screen time? Just to be clear, there is still no evidence that screens are actually beneficial for children under three, and most of the evidence for educational benefits in older children comes from studies of parents and children co-viewing screens together. Think: singing and dancing along, pausing to talk with your child about the content, interacting and responding as if the TV character is really talking to you (moms with a performing arts background will love these recommendations!). Co-viewing is wonderful for children's learning and development, and definitely shouldn't be dropped from your routine. However, one nice thing about introducing some solo screen time into your child's day is that you'll get the opportunity to spend a few minutes to yourself now and then! Screens certainly aren't the only way to practice independent play, but they do have the benefit of being mess-free, attention-focusing, and physically safe (unless your child is an acrobat who likes to climb the furniture while watching, which honestly isn't such a bad thing to offset the sedentary effects of TV). In most cases, this means an easy, breezy, hands-free 10-15 minutes for mom. What will you do with your time?

If you're like us, any 10- to 15-minute time increment is a convenient opportunity to finish a chore around the house. Washing dishes, prepping meals, folding laundry, even squeezing in a quick workout are all great activities to do while your child is focused on their age-appropriate tech activity! Or, you could choose our favorite way to spend free time, by brewing a hot cup of coffee or tea and reading a physical book or just taking a mindful moment to yourself. Even when they're playing with a tablet or watching TV, kids are learning from your actions and modeling how to pattern their own behaviors around self-care when they get older. Why not show them how it can be possible to relax without a screen?

Avoid Tantrums and Roll with the Punches

Technology is exciting! So exciting that it can sometimes lead to big emotions. What can you do if you are starting to get into a troubling behavior pattern of power struggles or meltdowns at home around devices? In this section, we offer some starting considerations to set a game plan in advance for avoiding overwhelming emotions associated with tech time.

Make sure devices are charged in advance, and set a clear expectation around time limits. You can use a visual like an hourglass or a familiar reference (like, "when I'm done making your lunch") to help children who are still learning about timekeeping. If your child's school uses technology, reach out to their teachers to see if there are any habits or routines they use that you can continue at home (e.g., "Apples up, Eyes up" when you want children to turn their iPad face down so the Apple logo faces up, and look at you). If tantrums tend to happen near the beginning of a session (e.g., frustration with wanting to negotiate for a longer play session), try to focus on how the device is making the child feel, rather than focusing on unwanted behavior. "I can see that this game that's supposed to help you feel happy and calm is making you even more upset. I never want to make you do something that's going to upset you for no reason, so let's think of something else we can do right now that would really help you feel great." In addition to communicating that device time is an option that can always be removed, this approach helps them reflect about their own mindset and mood when using tech, and de-emphasizes a narrative they may be forming that they are somehow a "bad" kid for feeling angry.

When you see a tantrum coming, stop and take a break to validate their emotions, name the feeling, and do any behavior habits (deep breaths, bear hug, quiet moment) that help your little one feel seen with their frustration. Playfulness is an amazing tool to break tension, change moods, and redirect overwhelming emotions. In general, tantrums are your child's way of seeking connection and comfort. Regular tantrums around screen or tech time might be a sign that they are not getting those needs met with a device. One strategy is to co-view with them for a few screen time sessions and make sure you fully understand the root of the cause of their tantrums. It may be as simple as teaching them how to pause and save work, or extending tech time by a minute or two to let them finish their last attempt at a challenging level of their game.

If your little one has a hard time with transitions, offering technology at a set time in your schedule might offer clarity and predictability to your child's routine. Keep devices (maybe even your own phone) away for when it's not tech time, and narrate aloud when tech is coming out to play. Set a timer that you always use for tech time to help ease them into or out of the digital experience. You might use more than one timer to close a session, such as a two- and one-minute warning, and a final timer with a different tone to signal that tech time is over, and it's time to put away the device. Make an agreement about tech time with children earlier in the day, and use the calm moments just before you hand over a device to remind your child of the expectations you both have agreed on–you can even have your child repeat the agreement and shake hands. For example, you might agree that they will turn off their device when the final alert goes off, and they should use the warning times to save their work, or make any last choices about fast-forwarding. Or, you might decide that the TV stays on for one episode of a show that you and your child choose together, and then it goes off before the next show starts. Whatever your agreements, be sure to keep these expectations consistent every time. Offer tech time during a slow-paced portion of the day to avoid adding undue stress when transitioning away from the screen. For example, if you know you'll need to remind your child repeatedly to stop the game, clean up the robot parts, or turn off and charge the tablet, you can avoid frustration by choosing unstructured downtime, like the half hour after breakfast, rather than a time-sensitive part of the day, like the half hour before soccer practice.

Remember that it is always okay to simply not offer technology as a choice to your young child. Screen time may be the norm for many families, but you are the only one who knows what your child needs to thrive. Any choice to step away from tech doesn't have to be permanent, and can give you a chance to reset and think clearly about how you want technology to serve you and what situations it will be best to offer it in your home. Some strategies to minimize tech use at home include:

- Turn off screens and remove them from bedrooms 30-60 minutes before bedtime.
- If your children have built a habit of expecting technology at a certain time in their routine, wait to offer tech until they specifically ask for it. You may be surprised how long they go. Alternatively, change up the daily routine and see if they notice that you omitted or shortened tech time that day.
- Don't rush to replace the batteries in digital toys (works best with really young kids)
- Talk with your kids about planning a social event, like a playdate, dinner with friends, or some other exciting outing that they will look forward to.
- Go outside!

From Digital Reward to Digital Resource

Real Mom Reflection

As our young children have gotten older, managing their screen time habits has become progressively more complicated—especially when they see that their friends and neighbors can "earn" screen time through good behavior, extra chores, or grades. Managing our children's access to digital devices has been a continuous learning process, and the temptation to offer tablet time or a TV show as a "carrot" for good behavior is powerful. It was even a strategy we briefly experimented with in our own household.

However, we soon came to realize that using screen time as a reward fundamentally misrepresents its role in our family's lives. This realization hit home with our elementary-aged son became overly performative in his attempt to earn extra Minecraft and Nintendo time. His "helpful" behavior was tinged with a frantic energy, and he was clearly motivated only by the digital prize. This transactional focus created more stress than it solved, teaching him to view every good deed as a negotiation for a reward. That's when we realized that while this system may work for our friends, it wasn't working for us.

By not linking screen time to behavior, we aim to teach our kids that it is an enjoyable resource, just like food or outdoor play, that needs to be balanced and managed responsibly—not a prize to be earned or withheld. To do this effectively, we've shifted our focus to establishing clear, predictable routines and time limits. Our kids know when they will have screen time and have choices about how to use this time. This shift has made conversations in our home less about "who's earned what" and more about "what's the schedule." Now that screen time is treated as a limited, scheduled part of our day (and no longer a reward), it has become normalized, and their intensity and desire for it has naturally diminished.

-Amanda Sullivan

Putting it all together: Connection, Content, and Context

In a policy brief issued in 2016, the Media Policy Project (based at the London School of Economics) argued that instead of focusing on how much screen time children are getting, families should ask themselves about "screen context (where, when and how digital media are accessed), content (what is being watched or used), and connections (whether and how relationships are facilitated or impeded)." We love this clear, thoughtful way to build your family's unique approach to screen time. Let's break these ideas down even more, to consider how you can practically use this recommendation in your day-to-day schedule:

- **Context:** This is where you get to adapt to your personal family values and culture. Do you think screen time should be earned through chores, or used as regular time to decompress? Will screens be available a few times a day? One day a week? Any time the child chooses, or specific times throughout the day? Are devices permitted in every room, or only certain spaces? What about screens at dinner time, or before bed? All of these questions pertain to your specific context, and there is no "right" answer to how, when, and where you permit your child to access screens. Just try to avoid making screen time too "special" or difficult to attain, to avoid children becoming fixated on the screen.

- **Content:** We've talked about the different kinds of content available to children elsewhere in this chapter, but it's important to remember that not all content available to children was designed with kids in mind. Try to offer options with educational and pro-social intentions behind them, and limit options that rely heavily on repetitive tactics like intense graphics, fast-reflex interactions, and auto-play features that make it hard to put down the device. Children's minds are in a tender, underdeveloped state and the research is still divided on the impact of the chemical rush that floods their brains during "mindless" addictive games. You can make the best choices for your family by choosing experiences that support the same kind of focused, regular-paced play that you love to see them engaged in off-screen.

- **Connection:** Every screen time session can allow children to connect or disconnect from those around them. Remember that children always need connection, but it doesn't necessarily have to happen during screen time. Think about the purpose of your tech break. Do you want to use screen time to bond with your child? Try co-viewing a favorite show, video-chatting a loved one, or asking them to teach you how to create a project with their favorite digital tool. This increases social connection. Do you need to get some chores done? Let your kids have some solo screen time to relax on their own, and then plan for a collaborative activity to reconnect when screen time is over. Maybe you want to use a mix of these. Think of your child's rhythms throughout the day to build habits that work for you and your family. Perhaps before lunch they are ready for interactive collaborative tech play, but when they become more reserved in the afternoons, they could benefit more from independent screen time.

How might it look to use the Context-Content-Connections framework in your daily planning? You could choose to use a screen time planner like the one below:

Sample Screen Time Planner

CHILD/AGE	SCREEN TIME ACTIVITY	CONTEXT	CONTENT	CONNECTION
3-year-old daughter	10 minutes of TV time	Before lunch- only after she helps me set the table	She can choose from her TV menu: Bluey, PBS Kids, or Ms. Rachel	We'll watch together
8-year-old son	25 minutes of video game time	During his sister's nap while I'm replying to emails	Minecraft - but he has to play in creative mode, not survival mode (I don't like the violence)	He can play alone, and afterwards he can write a sentence about his newest creation to share with the family at dinner

Will the screen time activity offer opportunities to foster connection with people they love? What will be the context for using their device? What content will they be allowed to choose from? Thinking about screen time in this way not only offers you a structure to keep screen time meaningful and engaging, it also helps to alleviate the anxiety that can come with handing a device to your child. You have a plan that you and your kids can feel good about, so you can lose the mom-guilt and move on to enjoying this fun stage of your child's digital play!

Screen Time Planner

Use the planner to think about how your kids will use their screen time.

Screen Time Planner

CHILD/AGE	SCREEN TIME ACTIVITY	CONTEXT	CONTENT	CONNECTION

CHAPTER 10
SCREEN-FREE TECHNOLOGY FOR YOUNG CHILDREN (IT EXISTS!)

What do you think of when you hear the word "technology?" Most people immediately think of digital (screen-based) technologies like computers, tablets, phones, and televisions.

Now, what do you think of and *feel* when you hear "children and technology?" Many people, especially moms, might immediately think of their kids watching their favorite television shows or playing their favorite tablet-based app. You may have a more complicated emotional response to this phrase. It may be associated with "mom guilt" about screen time or be associated with worries about your child being exposed to too much technology, despite all the benefits of certain types of media exposure discussed in the previous chapters.

Often what we visualize when we hear the phrase "children and technology" includes images of children sitting quietly with headphones on playing a game, or children passively watching a television show. Our idealized image of young children as noisy, playful, social, and messy usually isn't what we envision when we think of "children's technology" and this may be where a lot of feelings of mom guilt come into play when we think about our own children using technology from an early age.

It's true, children and technology usage often looks like quiet, passive, engagement with a screen. But what if we told you "children and technology" could also be:

- Social and collaborative
- Creative and open-ended
- Fun
- Educational
- Screen-free

Yes, you read that correctly, we said that technology could be screen-free. In this chapter, we are hoping we can change your conception of what "children's technology" looks like by providing examples of screen-free tools that can still support young children's understanding of high-tech concepts related to coding and other related topics we introduced in previous chapters on meaningful screen time and developmentally appropriate technologies for different ages.

So if you are ready to strengthen your tech-savvy mama skills (and be ultra-prepared with gift ideas for any upcoming birthdays or holidays) read on to learn about the different types of screen-free tools and resources available!

Computer Science Unplugged

Many moms of young children know that during these early years, kids are curious about the world around them and want to learn *how* things work and *why* they work the way they do. As young children grow older, they become more curious about how their favorite things like tablets and video games work. This becomes a wonderful opportunity to capitalize on our children's natural curiosity by teaching them about technology and computer science. Children can learn that their favorite apps and digital games all work because of something called code.

You may recall that back in Chapter 7, we talked about the many benefits of teaching coding and supporting computational thinking with young children. Unfortunately, despite the benefits, the complications of screen time and reliance on expensive digital devices present roadblocks to many parents (ourselves included!). Here is where the "unplugged" approach to computer science education has become a powerful movement over the past two decades! Within this movement, educators have recognized the value of integrating activities that do not require knowledge of computers or other technologies into the computer science curriculum. This unplugged (i.e. "tech-free") approach focuses on teaching programming concepts through puzzles, games, art, and more all without a computer, robot, or tablet. Originally used in schools and informal learning settings, unplugged computer science is now being explored at home with parents and caregivers through games, hands-on tools, and creative activities.

So where can you find a great list of unplugged activities? A quick Google search of a phrase like "unplugged coding activities for four-year-olds" is sure to yield a bunch of exciting content. You can also check out the original source of unplugged Computer Science content at csunplugged.org. This website contains a great collection of free learning activities that teach Computer Science through games, art projects, and puzzles that use cards, string, crayons, and even physical activities like running around. You can filter by age or topic to find activities, and you can even find a whole page dedicated to at-home activities.

Did You Know?

The original Computer Science Unplugged project was based at Canterbury University and has since been widely adopted internationally (translated into 12 languages) and it is also recommended in the Association for Computing Machinery (ACM) K-12 curriculum.
Get more information or access activities on the CS Unplugged website:

www.csunplugged.org

Find activities specially designed for at-home learning and play here:

www.csunplugged.org/en/at-home/

Board Games to Explore Coding

From within the CS-Unplugged movement, several coding-themed board games and card games have been growing in popularity over the past decade. Board games and card games are one of the easiest (and most fun!) ways to explore computer programming with young children at home. These games are the perfect addition to any family board game nights you may already have traditions built around in your home. Coding board games are also great for families with more than one child because they usually can be played with multiple players of mixed ages. Playing these games as a family can help younger children learn and understand the rules of the games faster than if they were to play by themselves.

Ready to start playing? Check out the table below for some specific board game suggestions, broken down by age. You can also simply type "coding board games for kids" into a search engine to find the newest offerings—there are more and more released each day!

Coding Board Games for Families

BOARD GAME	AGE RANGE	COST & DESCRIPTION
Robot Turtles	3+	$21 - Multiplayer board game with the goal of programming your turtle to navigate a maze to reach its jewel.
LittleCodr	4+	$13 - A card game where kids get to program their parents or friends to do crazy things by using simple action cards.
CoderMarz	6+	$34 - This unique board game teaches Mars facts, coding, and Artificial Intelligence concepts using fun gameplay.
Coding Farmers	7+	$14 - Children play the game with action cards in two ways: regular English and Java code. By playing the game several times, kids learn to connect their actions with written code.
Coding Master	8+	$20 - In Code Master, your avatar travels to an exotic world in search of power crystals. Along the way, you use programming logic to navigate the map.

Note. These are online prices based on the time when this book is being written. Cost and availability may fluctuate. We are not affiliated with these products in any way.

Screen-Free Programmable Robots

If you are looking to get even more high-tech with your playtime (but still avoid screens), there are a growing number of educational robotics products available for children as young as four to get started exploring computer science, engineering, and problem-solving. Research, including some of our own research conducted with the DevTech Research Group started out of Tufts University (now based out of Boston College), has shown that with simple robotics and programming languages, children as young as four can learn foundational engineering, programming, and robotics content. In addition to mastering this new content, we have also seen that playful learning with robotics can foster positive benefits for children's developing numeracy, literacy, and visual memory, and can even prompt collaboration and teamwork.

That all sounds great, right? On the other hand, you may also be thinking that robots sound expensive, fragile, or too high-tech to be truly playful and appropriate with rambunctious toddlers and young children at home. Affordability is something that only your family can decide, but you might be weighing the investment of a STEM tool against the potential educational outcome. Like anything else, it isn't always the cost, but the *purpose* of the toy that might make you decide it isn't the best choice for your family. Fortunately, there are a growing number of diverse and fun robot toys and kits out there. These range in price from fairly affordable (under $50) to a very big investment (over $100 and beyond). Let's break down the features you can expect to see at different price points and check out some examples of robot tools currently on the market for young children.

But first: what's the big difference between a cute toy robot that is simply a cute toy and a cute toy robot that is also designed as a STEM tool? The biggest difference is usually that an educational robot is one that your child will learn to code its actions as opposed to just playing with it. We break down the other characteristics of educational robotics tools in the table below.

Cute Robot Toy vs. STEM Learning Tool

CUTE ROBOT TOY	STEM LEARNING TOOL
<ul><li>Plushies</li><li>Wind-Up Toys</li><li>Simple remote-controlled toys</li><li>Does not allow for coding the robot (i.e., giving the robot a series of instructions)</li><li>Does not allow for building or personalizing the robot</li></ul>	<ul><li>Allows users to code their robot (i.e., giving the robot a series of instructions)</li><li>Allows users to problem solve when they don't get something right</li><li>Includes opportunities to build, engineer, or construct</li><li>May include opportunities to interact with sensors, sounds, or lights</li><li>Allows users to personalize and interact with the tool creatively</li></ul>

It may not come as a surprise that many of the best educational robots that allow for a variety of different STEM learning experiences come with a bigger price tag. For example, while a robot that allows you to simply practice giving instructions may run you around $50-$60, a robot that allows you to build, code, and decorate may be well over $100. Robots with extensive features, like interactive sensors, could be even more expensive. This can be daunting to parents, especially given the rapidly changing nature of technology and how quickly kids tend to lose interest in new toys! Keep a few tips in mind as you decide whether or not to invest in a new robot for your kids:

- **Find a way to try it first**—You want to invest in a tool that your child will enjoy using and is appropriately challenging for them. If possible, find a way to try out a new robot (or any big tech investment!) with your child to gauge their interest at a local library, makerspace, or children's museum before purchasing the product yourself. If you can't find a way to do this, read the refund and exchange information carefully before you buy!

- **Pick something they won't age out of quickly**—Many educational products for kids are tailored to a very specific and small age range. These may be cheaper in the short term but could mean you are ending up making more purchases of the same kind of tool in the long term. If possible, try to pick a robot that is simple enough for your child to start with while they are young, but has the ability to add more complexity as they get older. On the other hand, if you have multiple kids of different age ranges, it may be okay if your child outgrows tools quickly so that it can be passed along to a younger sibling!

- **Pick something with a high replayability factor**—The last thing any parent likes to see is an educational technology that they researched and invested hard-earned money on sitting to collect dust on the shelf after just a day or two of play. One way to prevent this is to invest in tools that have a high "replayability" factor. By this we mean, a tool or toy that children will want to engage with over and over again in new ways. This may be a robot that allows you to build it in different ways so that it looks different every time, a toy that is hard enough to challenge your child (without being SO hard that it makes them want to quit using it), or a toy that has unexpected surprises (noises, lights, or other features) that delight your child and make them want to use it again and again. There is no secret formula for replayability because every child is drawn to different things. Be sure to read as many product reviews from other parents as possible in order to learn more about the features of any product you are considering.

So what exactly does a robot for young children look like? Let's check out a few of our favorite examples!

Fisher Price Robots

At the time the two of us first began our professional journeys in educational technology, there were almost no coding or engineering toys out there for very young children (except of course, the ones we were working on developing!). Once the well-known American toy company Fisher Price (which specializes in educational toys for infants, toddlers, and preschoolers), entered the market with codable robots, it was safe to say that these types of tech tools were becoming part of the mainstream toy culture.

Code-a-Pillar, from Fisher Price, remains one of the most popular and well-known coding toys for young children. This robot caterpillar comes to life when it is assembled using its interchangeable segment pieces. Each segment represents a different action, sending the caterpillar in a different direction every time.

Think & Learn
Code-a-pillar

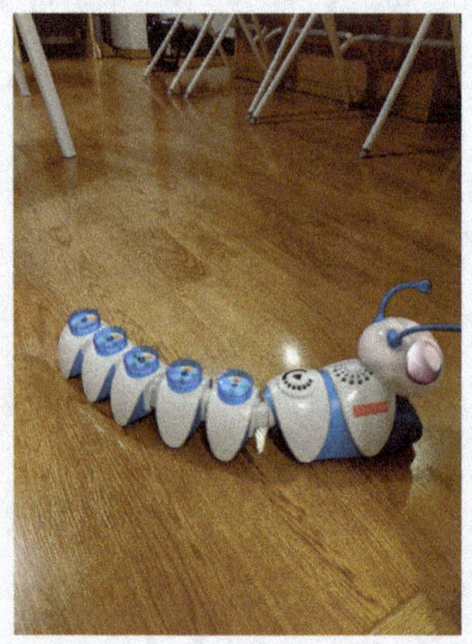

Another Fisher Price robot that delights our own kids is called Code 'n Learn KinderBot. KinderBot is an interactive robot (described as a "robot friend" in its product description) designed especially for preschoolers ages three to six years. This robot has a high replayability factor, with lots of different ways you can engage with it! At the most basic level, you can code Kinderbot to move using the directional arrows on the robot's head. But there are three different modes kids can explore: freely coding with the arrows, taking on challenges spoken to the kids by Kinderbot, or coding KinderBot to do fun things using a secret codebook.

Code 'n Learn
KinderBot

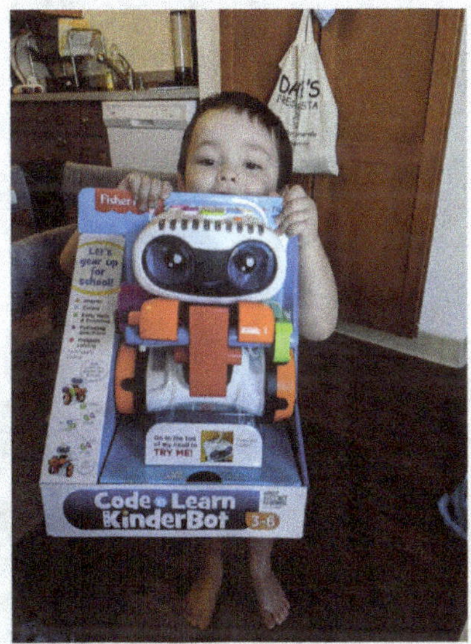

Learning Resources Robots

For the young child that loves adorable animals, the robots from Learning Resources may be a hit for exploring coding. With the Code 'n Go Mouse from Learning Resources, children can create a step-by-step path for the programmable robot mouse. The mouse lights up, makes sounds, and features two speeds along with colorful buttons to match coding cards for easy programming and sequencing. This mouse offers a variety of different ways to explore coding in a playful way, ranging from creating very simple paths to more complex mazes all outlined in an included Activity Guide.

For even more playful coding exploration, Learning Resources offers other adorable robots so that families can add more "coding critters" to their homes. Coding Critters Scamper & Sneaker are playful robot kittens designed to engage preschoolers with coding through a storybook context. Kids code along with their new pets' storybook adventure and help the curious Scamper and sneaky Sneaker have a playtime they'll never forget! Each storybook coding challenge encourages children to code and play in different ways in an imaginative story context, like coding a game of hide and seek. In addition to following along with the storybook's coding challenges, kids can also design and code their own games, stories, dances, and more!

Coding Critters
Scamper & Sneaker

BeeBot Robots

The BeeBot is a programmable floor robot developed for preschoolers. BeeBot, a friendly bumblebee-inspired robot with arrows on its back, can be coded to move in any straight direction (forward, backward, left, or right) and can be used with visual coding cards to plan and keep track of which buttons are pushed in which order. Although there is a suite of newer BeeBot tools including a bluetooth-enabled version (BlueBot) and an app, the original robot is simple, rechargeable (save on batteries!), and a great introductory tool for children ages three-plus to understand the correspondence of a coding command to the action of the robot.

BeeBot is the perfect example of a robot that is fantastic for a specific age band, and robust enough to be a favorite hand-me-down as children grow out of that narrow target range. Used as a math manipulative in schools around the world, the robot has been deeply researched for its educational benefits, and is a great starting point for more complex robots that can do more than simple movements. While our kids have found creative ways to keep the fun going with games like BeeBot treasure maps (code the robot to reach a certain point on a floor grid to reveal a surprise), the simple mechanics make it a great foundational toy whose appeal may not last once children master its four coding commands.

The KIBO Robotics Kit

We couldn't let this chapter go by without writing about a screen-free tool we (the authors) worked together at Tufts University for many years to help develop: KIBO. KIBO is a robotics kit designed specifically to playfully introduce young children (around ages four to eight years) to foundational engineering and programming concepts through unplugged play and activities. KIBO was created based on research by the Developmental Technologies Research Group at Tufts University (a lab now based out of Boston College) and made commercially available through funding from the National Science Foundation (NSF) and a successful Kickstarter campaign. KIBO is unique as compared to its counterparts, including the robots previously mentioned, because it engages children with both building with robotic parts (KIBO's hardware) and programming the robot to move with tangible programming blocks (KIBO's software). KIBO is designed based on years of child development research led by our mentor, Professor Marina Umaschi Bers, and is intended explicitly to meet the developmental needs of young children. The kit contains easy-to-connect construction materials including wheels, motors, a lightbulb, a variety of sensors, a wooden art platform, and more!

KIBO Robotics Kit

So how does KIBO work exactly? It is simple enough for any mom to feel like a coding expert! KIBO is programmed to move using a tangible programming language that consists of interlocking wooden programming blocks. Using this block language, children are able to master increasingly complex programming concepts such as repeat loops, conditional if/else statements, and nested loops (e.g., a repeat inside of another repeat). These wooden blocks resemble familiar early childhood manipulatives such as alphabet blocks and contain no embedded electronics or digital components. Instead, KIBO's main body has an embedded scanner that scans the barcodes on the programming blocks in order to "read" the program the child has created. Once the program has been scanned, it is saved on the robot instantly. No interaction with a computer, tablet, or phone is required!

Along with these robotic and programming components, the KIBO kit also contains art platforms that can be used for children to personalize their projects with crafts materials and foster STEAM integration. Unlike KIBO, many robots for young children come out of the box already decorated to look like a toy or creature. For example, the BeeBot is designed to look like a bumblebee. The Wonder Workshop robots Dash and Dot (robots programmed through a tablet application) are round and blue with a large eye in the center, resembling a friendly creature. KIBO on the other hand does not look like anything until the child places their imagination on it. It does not have a face and is constructed of neutral colors and wooden materials. Like an unsculpted wad of play-dough, KIBO looks and feels different each time your child uses it.

Picture Books to Explore Technology

If the thought of some of these robotics kits overwhelms you, think about starting with something as simple as a book—it doesn't get more unplugged than storytime! If you're wondering what books have to do with being a tech-savvy mom, remember that picture books offer a great way to start exploring and educating your little ones (and you!) on big topics like computer science, engineering, and more. Picture books also offer a great way to help counter stereotypes about who can and should pursue high-tech STEM fields when they present diverse characters from a range of backgrounds. From baby board books to longer informational texts, access to these books is the perfect way for parents to begin raising empowered digital citizens from an early age. Not sure where to start? Check out our "starter list" of picture books to get on your bookshelf!

STEM-Powered Book List
For Young Children

Start exploring high-tech topics in a very low-tech way... through picture books!

Here are a few suggestions that are perfect for bedtime stories or anytime read-alouds:

- Rosie Revere, Engineer by Andrea Beaty
- Ada Twist, Scientist by Andrea Beaty
- Robots, Robots Everywhere by Sue Fliess and Bob Staake
- How to Code a Sandcastle by Josh Funk
- How to Code a Rollercoaster by Josh Funk
- Baby Loves Coding by Ruth Spiro
- Baby Loves Structural Engineering by Ruth Spiro

Keeping Up with New Technologies

We could list every single technology (unplugged and screen-based) that are out there for young children at the time of writing this book but by the time it is published, half of them would no longer be available and there would be a whole new slate of tech tools being offered in its place. The goal of this chapter is not for you to learn the specifics of every unplugged tech tool or robot out there. Instead, it is to give you things to keep in mind when you are shopping for technologies for your children. We especially hope that after this chapter you have a new mindset on what children's technology can and should look like.

Digital Motherhood Self-Reflection #7

Here is a final self-check for you to keep in mind when considering a new technology purchase for your family (this checklist works for both unplugged and screen-based technology):

· Is there a high replayability factor beyond a few uses?

· Will my child be exposed to STEM learning content like coding or engineering?

· How quickly will my child grow out of this technology?

· Will it be FUN and ENGAGING for my child?

· Am I comfortable with the cost?

· Am I comfortable exploring this with my child and providing help/troubleshooting if they need it?

Real Mom Reflection

If you can answer "yes" to most of these questions, chances are you have found a great tool for your family—have fun learning and playing together!

CHAPTER 11

TECH-WISE TIPS FOR HOME AND TRAVEL

Let's start this chapter by painting a picture of different parenting scenes that may or may not feel familiar to you…

Scenario 1: A Road Trip

Your family has just embarked on a long-awaited road trip. Your seven-year-old son excitedly chatters about the adventures ahead, while your three-year-old daughter plays with a doll in her car seat. The trip starts off smoothly, and your son settles down to use his tablet. Although your daughter complains that she wants a turn with her brother's tablet, you remind her that she already used up her screen time this morning. The issue resolves itself when the tablet suddenly loses battery and your son's good mood deteriorates. You and your partner find yourselves navigating a chorus of requests and inquiries from the backseat. Soon, you have two crying children on your hands. You start digging into your bag of tricks, pulling out coloring books and snacks in the hopes of finding a distraction. Eventually, you pass back your phone and spend the next hour coordinating a delicate trade agreement between your two children that allows them to take turns sharing the phone. Your partner navigates the busy highway, and both of you begin counting down the miles to your destination.

Scenario 2: A Long-Distance Flight

Boarding the plane, you anticipate a smooth flight, timed perfectly to your baby daughter's naptime. Her five-year-old big sister quietly plays her favorite video game on the tablet, keeping her distracted while you get the baby settled. Unfortunately, the baby begins to fuss for no obvious reason just after takeoff. While you try to figure out what's wrong, your older daughter insists she needs to turn the volume up to compete with the loud crying, which further upsets the baby. Soon you're raising your voice and threatening to take away the video game, while both children lose control. You start to notice some raised eyebrows from the other passengers.

Scenario 3: A Summer Vacation Day

At home during a summer break from school, your family has planned a leisurely day together. You originally planned to spend the day at the pool, but as soon as you learn the pool is closed for a swim meet, your eight- and six-year-old sons fixate on watching cartoons on TV. You desperately try to engage them in other activities, but their attention remains glued to screens. Frustrated by the relentless reliance on technology, you decide it's time for a digital detox and suggest a family outing to the park, hoping to reclaim some quality time away from gadgets. The boys grudgingly agree, dragging their feet and complaining bitterly the whole time. You eventually make it to the park, but it takes 35 minutes of cajoling and demanding. By the time you arrive, you are exhausted and just want to sit on the bench scrolling on your phone, which causes both boys to complain and then beg you to let them take turns playing apps on your phone. In the end, you go home and let them watch their cartoon, feeling guilty and exhausted while you get ready to cook dinner.

Did any of these examples ring true for you? Or maybe they made you think of another tech struggle you often encounter with your family? We've all been there! In this chapter, we'll share some strategies for keeping tech use manageable and stress-free, whether you're building a stable everyday routine at home, or you're on the road, away from your home-base structures that keep things running smoothly.

Rhythms at home: Making tech time work for you

Let's first tackle the home routine. We understand that this is easier said than done. Between mountains of laundry and dirty dishes, special school events, afterschool extracurriculars, holidays, illnesses, and developmental "phases" that throw a wrench in the entire day (school refusal, anyone?), it might feel impossible to find a daily rhythm at your house. As a mom, you probably know that running your household is a bit like maintaining a very delicate ecosystem, and it only takes one runny nose or last-minute homework assignment to shift the week in a chaotic direction. We've been there, and we know that embracing the chaos is part of the journey of motherhood. But if you feel like technology is adding to the chaos, we do have some suggestions to help your tech/life balance.

First, as with all routines, we opt for a habit over a schedule. Most families we know don't sit down to dinner at exactly the same time every night before a hard-and-fast bedtime curfew. More likely, you have built a reliable series of activities to transition into and out of dinner and bedtime. These transitions may have been very intentional, or they may just be what naturally works for you and your kids. Either way, humans are creatures of habit and we like to have a familiar set of habits that we can expect every day.

When it comes to creating a tech habit, the approach is pretty much the same no matter what kind of shift you want to make in your family. Building a habit is easy: repetition. Changing a habit is slightly harder, but all it requires is changing your environment to make it easy to practice a new, preferred habit, and hard to repeat the old one. For example, let's say you've fallen into a comfortable routine of using the iPad before bed, but lately it's started to encroach on bedtime. It may feel impossible to stop your kids from asking for iPad time after they change into their pajamas, but after a few days of reminding them that the iPad gets locked in your home office every night before dinnertime so it can charge overnight, you might be surprised how quickly they learn to ask for iPad time earlier in the evening–or develop a new bedtime routine using a screen-free robotics or engineering kit!

If you're thinking of starting or changing the tech habits in your house, the most important thing is to think of what level of technology you would feel comfortable with in your home, and use that as your guiding compass. Think of your end goal, and what you want your ideal routine to look and feel like. You can dream big, but be realistic about where your family is starting. Some ideas for building reliable tech habits at home include:

- *Create a menu of acceptable tech choices*—A written or visual list with pictures of different apps, games, robots, shows, etc., that kids can choose from during designated tech times can smooth the transition into tech time. These menus might look different for different times of day, different devices, or different days of the week. You can have as many menus as you like, but having a short list of choices helps *you* by limiting the back-and-forth negotiation about what's "allowed" right now, and helps *your child* by focusing their options so they spend less time choosing and more time playing.

- *Make technology experiences shared and communal*—Some families find it really helpful to offer a shared TV or computer for all tech time instead of offering a personal device, and introducing set times throughout the day or week when the shared device is "open for business". That way, you don't have to take the device away or worry about hiding it when it's not in use, because the whole family is accountable to the shared screen. You can also rely a bit on social pressure from family members to keep the balance of how often and what kind of tech activities are happening. We love this solution, because it also involves the whole family in the troubleshooting, brainstorming, and creative aspects of playing with a technology tool or digital design app.

- *Create tech-free zones*—This could be times of day, locations in the house, or between clear event boundaries (e.g., before homework, after dinner). The benefit of this approach is that it creates an obvious rule about when it's okay to use technology, shifting a request for tech time from a negotiation with you to a simple check to see if the rules are being met to allow tech time right now.

- *Build predictable tech time into your day*—Sort of the opposite of a tech-free zone, which enforces a boundary around when *not* to use tech, this approach builds tech time into your regular routine when children know they can predictably expect to be allowed to use technology. The thinking behind this approach is that if you are consistent with when you offer tech, such as during a younger sibling's nap, your child will come to naturally associate technology or screens with that part of their day, and it will make sense that those tools don't belong in other parts of the routine (like family meals and busy outings). For some kids, this can have the effect of making tech time feel less restrictive, and alleviate struggles that might be happening due to inconsistent rules around technology access. If you want to start this approach, we recommend aligning tech time with other rock-solid habits that your family already practices a lot (e.g., while you're making lunch, or after they're done cleaning up their room), to help keep that feeling of consistency before and after the technology break.

At the end of the day, remember that whenever your child leaves the house, to go on vacation, heading off to camp, or even going to school, they are heading into spaces where the rules are out of your control—technology-related or otherwise. The best we can do as moms is ensure that our homes are as consistent and empowering as possible for all our family members, so that our kids always have a safe home base to return to.

Tips and Tricks for Taking Tech on the Road

Gearing up for a big trip? This section will help you plan ahead so you can make the most of your travel time and let technology help you and your kids relax!

The best way to feel prepared about bringing tech tools on your trip is to have a plan going in. If you're used to packing for your kids and thinking ahead about their clothing and medical needs, then hopefully it doesn't feel too daunting to add a few minutes to plan for their tech needs as well. Is there a favorite show that always keeps your kids in a good mood? Queue those up on your tablet or computer first before a long drive or flight. In addition to feeling good about offering them a show they're familiar with, you can also track exactly how much content they have available and how long it will last. Use airplane mode to test favorite apps and games BEFORE you travel, to learn whether they require internet. If you aren't able to play their favorite app without WiFi, don't worry! With enough lead time, you can introduce new apps that don't require internet access. Alternatively, if no other app will do and you don't mind a paywall, many airlines and railways let you call ahead to check if your specific trip allows en route WiFi purchases, or you can add a hotspot to your phone and bring internet anywhere you go with cell service. Don't forget to pack a travel charger and cables for any devices. You don't want to be stuck without battery power when your children's energy is already drained!

If you have kids who love to listen to music, podcasts, or sound effects while they use their device, get them a pair of wireless headphones. This is a fun way to keep the noise level contained, while also inviting kids to personalize their style and get involved in their own device maintenance. Older kids are more likely to actually wear the headphones if they try them on and pick them out themselves, and we've noticed that younger kids seem drawn to over-ear headphones with plush cushioning on the ear. You can also use this trick to personalize their whole device with a fun charging cable in their favorite color, or a device case they can choose or decorate themselves. Customizations like this imbue ownership and can help your child to remember to pack their tech accessories at the end of your trip. If headphones don't end up working for your little one, you can always lean in and play their favorite songs on the radio or through your phone. The music may be repetitive, but it could foster a love and shared family culture of music that can help your child feel a strong sense of belonging and safety on car trips–and if they're like our little ones, playing music aloud encourages them to find their beautiful sing-along voice!

How much is too much tech time when you're traveling?

Just as travel can mean an upset to your daily routines around sleep, meals, and downtime, expect your tech time to look different on the road. Cramped spaces, sleepless hours, new experiences, and unfamiliar foods, sights, smells, and sounds can all lead to a lower threshold for stimulation than your kids usually display. Here, technology can help by redirecting some of their energy from feeling lost, bored, or overwhelmed to focusing on something engaging.

Be reasonable about the demands of travel. If you keep a very strict screen diet at home, recognize that travel is going to throw that schedule into the air. Any trip where kids must stay seated and generally not act their age for longer than two hours might be a good time to make a special exception for device-based play and entertainment. If you have children of multiple ages and you're trying to limit screen time for one but not the others, consider screen-free tech for your younger. For example, battery-operated busy boards with buttons, lights, and switches (always avoid choking hazards) give the experience of something technological, without the attention-fatiguing effect of a screen. Or, you can offer more of the tech breaks that you already feel comfortable with, like extra whole-family video calls to loved ones.

You can also try to provide reasonable travel-based boundaries around screen time. If your children love an app that requires internet, you could limit using it to anywhere with *free* internet during your trip, like airplane terminals and highway rest stops. Be sure to offer engaging alternatives, such as coloring books, puzzles, handheld games like etch-a-sketch, or fidget toys, during a travel stretch without WiFi. If you are worried about offering more technology than they're used to, don't be too hard on yourself. Travel takes a lot out of everyone, and kids deserve to relax, too. If you can offer green time outside to offset screen time as often as you can when you aren't actively en route. Take breaks, hydrate and eat snacks, and offer plenty of opportunities for off-screen connection like talking about what they're doing on their devices, or discussing what's going on around them in the airport, train station, or on the highway.

Letting Someone Else Drive: How to hand off tech management to a trusted caregiver

What if you're the one traveling, but your kids will be staying with a babysitter while you're away? You may be surprised that our recommendations aren't very different from travel: loosen the reins and let the kids live a little. When their reliable parental figures are unexpectedly gone for even a portion of a day, it can throw a young child's whole world into a bit of confusion. Leaving the kids home while you spend a few nights out of town can be a great opportunity for young children to feel a bit of independence and freedom from the regular routine, and to build a bond with another trusted adult. But first, they'll need a baseline level of comfort and security to enjoy those feelings of empowerment. The best thing you can do as a parent is to help support your child's happiness around the new person who will be watching them, even if that means their relationship dynamic and daily routine look different from what they may be used to when you're home. That doesn't mean you have to throw all rules out the window, but maybe decide where you can let some flexibility into their day, and where you want to draw some simple, firm lines that the babysitter should know not to cross.

You can help caregivers by leaving many options (both digital and tech-free) for approved play choices, to lessen the chance that they will run out of ideas and offer something that you definitely don't want to allow in your absence. The same visual and concrete information that helps kids can help adults, so you might leave a printed list with pictures of acceptable TV shows and software programs, and clear instructions for tech routines or time limits. Finally, think of the little logistics that might get in the way of a helpful plan. Look at your home through the eyes of a stranger, and think what might not be obvious to them. Things like leaving a note to explain the visual timer you always use for screen time can help eliminate confusion for house guests, who might also be navigating your home for the first time and trying to find where you keep the water glasses and spare toilet paper.

If you have a clear boundary that you know you'd feel uncomfortable with your kids crossing, say what you mean. Don't offer choices that you really hope they don't resort to, because these adults don't know your kids like you do and might offer "last resorts" too quickly. If possible, try to arrange your home to make the choices easy for everyone to stick to. For example, you can leave devices in airplane mode and ask a sitter to only use internet-optional apps and games. If it's possible, make yourself available for questions from the sitter at certain times of your trip. Most importantly, evaluate your situation for safety. You might feel concerned that your babysitter will do things so differently from usual that your kids' routine will be chaotic for weeks, but is that really such a terrible price to pay for their safety and wellbeing in your absence? Ensure that the adult in charge never shares inappropriate, violent, or harmful content, and then ask yourself what level of change from their normal tech habits is acceptable to you, and what tech choices you would really caution against to help the kids maintain a positive relationship and avoid conflicts with the sitter. If your children are thriving when you get home and are just a little wired at bedtime, or extra-persistent about asking for more screen time for a while, hopefully that's an acceptable trade-off for the kid-free vacation you just took to rest and reset. If you aren't sure how to draw the line between freedom and structure for your kids, you can treat your next getaway as a chance to collect some data. Ask sitters to note down and share with you when they started and stopped device time and what they worked on. This gives you some baseline information for the next trip you take, and some talking points if you want to have a conversation with your kids about how they spent their time while you were away.

Finally, be patient with this new experience. Setting one or two firm and easy-to-remember boundaries with a caregiver is a reasonable expectation, but remember that this is likely a new experience for them and for the kids. Hopefully, your well-intentioned caregiver will do what they think they need to do to keep your children happy, at peace, and comforted in your absence. If you still aren't confident about tech time while you're away, keep in mind that your kids are going to forge their own unique relationships with other caregivers (especially if they are family members who will probably play a large role in their lives), and the best you can do is hope that you raised them well enough to make good choices when you aren't around.

Detoxing from Tech Overexposure on Vacation

Real Mom Reflection

"The summer before preschool, we had scheduled a lot of travel time on our weeks off. One unexpected side effect was that we noticed our daughter getting a lot more demanding about screen time. Even though we kept our usual minimalist approach to screens at home and spent plenty of time outside, she was getting used to TV shows and movie nights at grandparents' houses and on vacation—places where the 'rules' felt out of our hands.

After thinking it through, we realized the real problem wasn't the occasional extra screen time at grandma's. It was that we weren't being consistent with screen boundaries once we returned home. So, we tried something new: we bought a pack of shiny plastic pool "gems" (which she adored) and used them as a little family currency system. She could earn gems by doing small chores—helping with potty training, setting the table, feeding the dog, or just being extra kind and helpful. Some days she got 2–4 gems, some days none.

Gems could be traded for rewards, and one option was TV time: one gem equaled about 5–8 minutes of a parent-approved show or video. Usually, she picked a short episode of Bluey. We were never 'sticker chart' parents and felt a little weird making TV a reward, but we wanted to experiment with the idea that the root of the behavior issue was a perceived lack of control (not necessarily wanting the TV itself). The turnaround was instant. What she loved most wasn't the shows. It was having some say over when and how she got screen time.

We were surprised that very quickly during the back-to-school season, the gem system faded. Once we were back in the swing of a regular routine with scheduled co-viewing and occasional family movie nights, she didn't bother to work for gems or even pester us for extra TV time. The lessons that TV comes after chores and has a clear end time stuck, and we all learned that what mattered most wasn't the gems or even the shows—it was the feeling that we were all on the same team. Instead of fighting over screens, we had a system that gave her some control and gave us consistency. In the midst of a busy vacation season filled with days that felt irregular and inconsistent for our daughter, that shift turned TV time from a power struggle into a temporary family routine we could all live with."

– Amanda Strawhacker

Rewriting the Script for Common Travel Scenarios

We started this chapter by thinking of common challenges that parents face when traveling. Let's revisit each scene, putting our technology tips in action to change the outcomes.

Scenario 1: A Road Trip

Your family has just embarked on a long-awaited road trip. Your seven-year-old son excitedly chatters about the adventures ahead, while your three-year-old daughter plays with a doll in her car seat. The trip starts off smoothly, and your son settles down to use his tablet. When your daughter complains that she wants a turn with her brother's tablet, you pass back her screen-free tech toy, a busy board with light-up LEDs and speaker sounds. Your son's tablet suddenly loses battery, but before his good mood deteriorates, you hand him the portable charger you always pack, and he connects it to his device. Thirty minutes later, you remind your son to pause screen time for a snack break, and after a couple more toy distractions and a phone call through the car-speaker to grandma, both kids drift to sleep in their carseats. You and your partner peacefully chat about the plans for the coming vacation, and both of you begin counting down the miles to your destination.

Scenario 2: A Long-Distance Flight

Boarding the plane, you anticipate a smooth flight, timed perfectly to your baby daughter's naptime. Her five-year-old big sister quietly plays her favorite video game on the tablet, keeping her distracted while you get the baby settled. Unfortunately, the baby begins to fuss for no obvious reason just after takeoff. While you try to figure out what's wrong, your older daughter puts on her favorite cat-ear headphones that she picked for her birthday, and you focus on feeding the baby. Soon your younger daughter is fast asleep on your chest. You watch while your older daughter plays her game, and smile quietly with her when she beats a tricky level.

Scenario 3: A Summer Vacation Day

At home during a summer break from school, your family has planned a leisurely day together. You originally planned to spend the day at the pool, but as soon as you learn the pool is closed for a swim meet, your eight- and six-year-old sons fixate on watching cartoons on TV. You have been frustrated by their reliance on technology during break, so you decide it's a great opportunity to talk together about some new agreements around technology time during summer vacation. You listen to their requests respectfully, and share your own ideas about what's fair. After agreeing on a structure of "screen minutes" earned through off-screen activities that keeps all three of you happy, you suggest an outing to the park to begin earning their first screen break. The boys excitedly agree, calculating how long they'll have to play at the park to earn a full episode of their favorite cartoon. By the time you arrive, they are in such a great mood you end up staying an extra half hour beyond their original goal. At home, they are tuckered out and collapse onto the couch to watch the cartoon they "earned," while you take some time to yourself to prepare dinner and listen to an audiobook.

Do these examples feel more attainable to you, after reading through the chapter? We hope you noticed that some things aren't possible to avoid—a fussy baby, older kids who beg for screen time, and kids who get bored on a car ride are all a part of the joys of parenting. But by building a game plan ahead of time, you can make technology work *for* you, not against you, to maintain family harmony. With the right boundaries that meet your family's unique needs, you can add technology to your tool kit of entertainment and distraction options, and keep your family routine flowing smoothly even in the face of unexpected vacation and travel changes.

Take a moment to think of your own home or travel tech scenario that you wish you could change. What's the most frustrating part for you, and for your family? Do any of the tips in this chapter give you a new idea about how to approach the challenge? Whether you are planning your next big trip (with or without the kids), or wishing for a different structure at home, we hope this chapter has given you some ideas that help you feel more confident to face the tech time challenges of a changed-up schedule or trip away from home.

PARENTING IN THE AGE OF AI

We hope after reading this book you have come to see just how powerful we, as moms, are in raising the next generation of innovators, technologists, and computer programmers, regardless of our own backgrounds. We play an immense role in ensuring that our children grow into informed, responsible, and caring community members of the ever-evolving digital environment in which they are immersed. Yet, we recognize that with the knowledge and ideas presented in this book come a whole new set of challenges placed firmly on the shoulders of moms. We know you already have your hands full mastering night feedings, teaching your child to ride a balance bike, or supporting your emerging reader to master a new list of sight words. It might feel like too much to think about navigating technology and screen time in a new way when all of this other "day-to-day" mom-survival stuff is at the top of the list. We get it, and there are many days when we feel exactly the same way. Simply by being reflective about the topics we have explored in this book and considering what works best for your family, we hope you know you are making a positive impact in the way you approach caring for and cultivating knowledge in your children.

Artificial Intelligence (AI) and All to Come

This book has explored strategies and examples for dealing with apps, digital media, and screen-based games that were once novel and not thought of in the same sentence as toddlers or young children. Coding and robotics for toddlers, preschools, and early elementary students, for example, have only become widely mainstream in the last decade. While we have drawn on as many examples of current trends and tech from our own research and work in the field of educational technology and media, we recognize that it is impossible to cover ALL of the types of technologies out there impacting mothers, families, babies, and children (it would be impossible for any book to!). And even if we could, the tools, apps, and technologies here today may be extinct tomorrow with something completely new in its place. Our hope is that the strategies, resources, and moments of reflection prompted by this book will be universal with any changes in the digital landscape to come.

Just in the course of researching and writing this book, hot topics like Artificial Intelligence (AI) have become a growing concern for parents and educators. As AI becomes more prominent, educators and caregivers are grappling with the best way to empower children to harness its potential in safe and appropriate ways. We know that AI is a powerful tool that can be used for good, for bad, and everything in between (just like all technological innovations). We also know that it is important to teach young children how to use AI responsibly and to be aware of the potential dangers. This starts with co-engagement between a parent or caregiver alongside your child, once you think it is developmentally appropriate. We would caution you not to wait too long though! While your preschooler or Kindergartener might not be quite ready for co-engagement and discussion, your elementary students probably are. New research from the National 4-H Council's survey "Youth AI Use & Understanding" shows that over 70% of kids between the ages of nine to seventeen who use AI would like adults, such as a teacher or parent, to help them learn how to use AI tools correctly and confidently.

Learning to engage with AI or any of the other digital innovations to come may seem daunting. It can help to think that the same principles of digital literacy apply to any and all new additions to the digital landscape. By honing your children's digital literacy skills, they will be prepared to navigate the online and technological world in thoughtful and responsible ways. Digital literacy is

NOT about knowing how to use particular devices. It is about thinking critically and making responsible choices when using technology. It is also about having the skill sets to engage with digital technology in creative ways, not just as *consumers* of digital content, but as *creators* of it. Evaluators of it. Critics of it. And eventually, leaders in the new industries that rely on digital technology and harness it in ways we cannot even imagine right now. While fostering these skill sets may not be as easily done as said, the approach boils down to a few approaches and practices:

- **Build your own digital literacy**—Explore the app/tool/game you want to talk to your child about on your own to build your own knowledge and confidence. It is okay not to be an expert, but a basic understanding (at least of any safety or privacy concerns).

- **Co-engage with your child**—Engage alongside your child whenever they are using a game or technology that is new for you/them.

- **Model your own critical thinking**—Model how you make smart decisions for yourself when engaging with technology. Maybe some of your social media profiles are private. Maybe you change your passwords regularly to make these conscious choices visible to your young children so that they can make smart decisions too.

- **Talk, talk, talk**—Creating an open dialogue about digital experiences now will be the basis of an open dialogue when your young children are tweens and teenagers. Make sure you are a safe place for them to come to with questions or concerns. Beginning when your children are very young and just starting to engage with digital media, you can tell them things like "if you ever see a video or picture that is scary or upsetting you can always come tell me and we can talk about it. I won't ever be upset with you for that." Remember that even if YOU take every digital precaution with your child, that does NOT mean that they may not be exposed on another child's tablet, phone, or computer. This is normal exploration and experimentation, and the most important thing is creating an open dialogue where your child feels comfortable talking to you about it so you can ensure they have as safe of experiences online as possible.

- **Revisit and review our Digital Motherhood Self Reflections—**Look back at your responses to any of the Self Reflection activities throughout this book. After reading this far, have any of your attitudes or questions changed? How do you view your personal and family technology choices, and what new practices have you tested, or are you considering? The fact that you're being intentional and mindful at this point in your parenting journey is a positive thing to celebrate!

Last Thoughts

It is indisputable that moms today must navigate new digital technologies, new worries, and new fears that past generations of mothers did not face. But it is also true that you, the moms of today, are resilient and undoubtedly more self-reflective and intentional about your parenting than any generation past. We are making mistakes, learning from our mistakes, and doing the best we can for our children in an intentional way each and every day. We are not perfect but we can and should look to our children with peace and confidence that no matter how technology and society change and evolve, new issues of morality or global crises that emerge, they are the generation of humans who will change the world for the better. And we are the moms with the privilege of raising them.

ACKNOWLEDGMENTS

This book was born from many late-night conversations, countless texts, and the shared experiences of navigating life as researchers, educators, and—most importantly—moms in the digital age. Writing it together has felt like an extension of those conversations, a way to process and share what we've learned (and are still learning) about raising children, building careers, and finding joy in both.

We are deeply grateful to our mentors and role models who have guided us throughout our professional journeys. In particular, thank you to Dr. Marina Umaschi Bers for opening our eyes to the power of playful learning and for believing in our work from the very beginning. Thank you to Dr. Julie Dobrow for your generosity, your encouragement, and your example of mentorship and leadership grounded in empathy and curiosity. You have shaped the way we both approach teaching, research, and even parenting.

To our publisher, Diane Windsor—thank you for believing in this project before it had a name, for supporting both of us through this book, and for being a steady voice of encouragement. You've been with us through every milestone and pivot, and your faith in this project made it possible.

We also want to offer special thanks to the people who lent their time and wisdom to this project. To our friend and longtime collaborator Angie Kalthoff: your thoughtful edits and peer feedback were invaluable in helping us clarify our message and keep the book accessible and engaging. And to Wendy Strawhacker: thank you for bringing a techie grandma's lens and a generational perspective to your generous read-through—your insights made this book richer and more relatable for parents of all kinds.

Beyond the team of people working directly on this book, we want to recognize the schools, teachers, parents, and—most of all—children who have shaped our understanding of what it means to grow up and parent in a digital age. Your stories, questions, and creativity inspired much of what appears in these pages. Thank you for letting us learn from you.

This project was built on the love, support, and patience of our families. Thank you to our immediate families—Adam, Sidney, and Sienna; and Jeff, June, and baby boy Soyster (who as of the writing of this book is due to come into this world the same year as this book!)—and our extended families for cheering us on, reviewing countless revisions and updates, keeping us grounded, and reminding us every day why these questions about parenting in a digital age matter so deeply. We also both gratefully thank all the moms who shared stories, insights, and solidarity along the way, and the lifelong friends who offered child-wrangling, coffee, and pep talks during the writing process—it truly does take a village.

Finally, we want to thank each other. Writing a book is a marathon and doing it as a team makes it infinitely more joyful and more meaningful. We are grateful to have found a "work wife" in one another—a co-conspirator, a friend, and a thought partner for all things playful, digital, and human. This book is as much a reflection of our friendship as it is of our research.

REFERENCES

Introduction:

American Academy of Pediatrics. (2016). Media and young minds. Pediatrics, 138(5), e20162591. https://doi.org/10.1542/peds.2016-2591

Auxier, B., Anderson, M., Perrin, A., & Turner, E. (2020). *Parenting children in the age of screens.* Pew Research Center. https://www.pewresearch.org/internet/2020/07/28/parenting-children-in-the-age-of-screens/

Bright Horizons Family Solutions, Inc. (2025, March 19). *A majority of parents admit even youngest kids need a "digital detox"* [Press release]. https://investors.brighthorizons.com/news-releases/news-release-details/majority-parents-admit-even-youngest-kids-need-digital-detox

Common Sense Media. (2018). *Common Sense Media/SurveyMonkey YouTube poll topline.* https://www.commonsensemedia.org/sites/default/files/research/report/commonsense-surveymonkey-youtube-topline_1.pdf

Parker, K., Horowitz, J. M., & Rohal, M. (2015). *Raising kids and running a household: How working parents share the load.* Pew Research Center. https://www.pewresearch.org/social-trends/2015/11/04/raising-kids-and-running-a-household-how-working-parents-share-the-load/

Rideout, V. (2017). *The Common Sense census: Media use by kids age zero to eight.* Common Sense Media. https://www.commonsensemedia.org/research/the-common-sense-census-media-use-by-kids-age-zero-to-eight-2017

Chapter 1:

American Pregnancy Association. (2021). Signs of miscarriage. https://americanpregnancy.org/getting-pregnant/pregnancy-loss/signs-of-miscarriage/

Bardos, J., Hercz, D., Friedenthal, J., Missmer, S. A., & Williams, Z. (2015). A national survey on public perceptions of miscarriage. Obstetrics & Gynecology, 125(6), 1313–1320. https://doi.org/10.1097/AOG.0000000000000859

Branum, A. M., & Ahrens, K. A. (2017). Trends in timing of pregnancy awareness among U.S. women. Maternal and Child Health Journal, 21(4), 715–726. https://doi.org/10.1007/s10995-016-2155-1

Bryant, A. G., Narasimhan, S., Bryant-Comstock, K., & Levi, E. E. (2014). Crisis pregnancy center websites: Information, misinformation and disinformation. Contraception, 90(6), 601–605. https://doi.org/10.1016/j.contraception.2014.07.003

Centers for Disease Control and Prevention (2022). Reproductive Health: Infertility FAQs. https://www.cdc.gov/reproductive-health/infertility-faq/?CDC_AAref_Val=https://www.cdc.gov/reproductivehealth/infertility/

Centers for Disease Control and Prevention. (2022). Depression among women. https://www.cdc.gov/reproductive-health/depression/?CDC_AAref_Val=https://www.cdc.gov/reproductivehealth/depression/

Cesare, N., Oladeji, O., Ferryman, K., Wijaya, D., Hendricks-Muñoz, K. D., Ward, A., & Nsoesie, E. O. (2020). Discussions of miscarriage and preterm births on Twitter. *Paediatric and Perinatal Epidemiology, 34*(5), 544–552. https://doi.org/10.1111/ppe.12622

Coyne, S. M., McDaniel, B. T., & Stockdale, L. A. (2017). "Do you dare to compare?" Associations between maternal social comparisons on social networking sites and parenting, mental health, and romantic relationship outcomes. *Computers in Human Behavior, 70,* 335–340. https://doi.org/10.1016/j.chb.2016.12.081

Dolan, B. (2013, February 14). Report finds pregnancy apps more popular than fitness apps. MobiHealthNews. (Archived)

Frid, G., Bogaert, K., & Chen, K. T. (2021). Mobile health apps for pregnant women: Systematic search, evaluation, and analysis of features. *Journal of Medical Internet Research, 23*(10), e25667. https://doi.org/10.2196/25667

Jankowicz, N. (2022). The internet is failing moms-to-be. *WIRED.* https://www.wired.com/story/pregnancy-apps-disinformation/

Jayaseelan, R., Pichandy, C., & Rushandramani, D. (2015). Usage of smartphone apps by women on their maternal life. *Research Journal of Science and Technology, 7*(3), 158–162.

Lee, Y., & Moon, M. (2016). Utilization and content evaluation of mobile applications for pregnancy, birth, and child care. *Healthcare Informatics Research, 22*(2), 73–80. https://doi.org/10.4258/hir.2016.22.2.73

Lima-Pereira, P., Bermúdez-Tamayo, C., & Jasienska, G. (2012). Use of the internet as a source of health information amongst participants of antenatal classes. *Journal of Clinical Nursing, 21*(3–4), 322–330. https://doi.org/10.1111/j.1365-2702.2011.03910.x

March of Dimes. (2022). *Baby blues after pregnancy.* https://www.marchofdimes.org/find-support/topics/postpartum/baby-blues-after-pregnancy

Sayakhot, P., & Carolan-Olah, M. (2016). Internet use by pregnant women seeking pregnancy-related information: A systematic review. *BMC Pregnancy and Childbirth, 16*(1), 65. https://doi.org/10.1186/s12884-016-0856-5

Sormunen, T., Karlgren, K., Aanesen, A., Fossum, B., & Westerbotn, M. (2020). The role of social media for persons affected by infertility. *BMC Women's Health, 20*(1), 112. https://doi.org/10.1186/s12905-020-00964-0

Chapter 2:

Auxier, B., Anderson, M., Perrin, A., & Turner, E. (2020). *Parents' attitudes—and experiences—related to digital technology.* Pew Research Center. https://www.pewresearch.org/internet/2020/07/28/parents-attitudes-and-experiences-related-to-digital-technology/

Auxier, B., & Anderson, M. (2021). *Social media use in 2021.* Pew Research Center. https://www.pewresearch.org/internet/2021/04/07/social-media-use-in-2021/

Gilkerson, J., & Richards, J. A. (2009). *The power of talk: Impact of adult talk, conversational turns, and TV during the critical 0–4 years of child development* (LENA Foundation Technical Report ITR-01-2). LENA Foundation. https://www.lena.org/wp-content/uploads/2016/07/LTR-01-2_PowerOfTalk.pdf

Hill, D. L. (2016). Why to avoid TV for infants & toddlers. *HealthyChildren.org.* https://www.healthychildren.org/English/family-life/Media/Pages/Why-to-Avoid-TV-Before-Age-2.aspx

Kirkorian, H. L., Pempek, T. A., Murphy, L. A., Schmidt, M. E., & Anderson, D. R. (2009). The impact of background television on parent–child interaction. *Child Development, 80*(5), 1350–1359. https://doi.org/10.1111/j.1467-8624.2009.01337.x

Mohd Shukri, N. H., Wells, J., Eaton, S., Mukhtar, F., Petelin, A., Jenko-Pražnikar, Z., & Fewtrell, M. (2019). Randomized controlled trial investigating the effects of a breastfeeding relaxation intervention on maternal psychological state, breast milk outcomes, and infant behavior and growth. *The American Journal of Clinical Nutrition, 110*(1), 121–130. https://doi.org/10.1093/ajcn/nqz033

Pempek, T. A., Kirkorian, H. L., & Anderson, D. R. (2014). The effects of background television on the quantity and quality of child-directed speech by parents. *Journal of Children and Media, 8*(3), 211–222. https://doi.org/10.1080/17482798.2014.920715

Schaeffer, K. (2021). 7 facts about Americans and Instagram. *Pew Research Center.* https://www.pewresearch.org/fact-tank/2021/10/07/7-facts-about-americans-and-instagram/

Schmidt, M. E., Pempek, T. A., Kirkorian, H. L., Lund, A. F., & Anderson, D. R. (2008). The effects of background television on the toy play behaviors of very young children. *Child Development, 79*(4), 1137–1151. https://doi.org/10.1111/j.1467-8624.2008.01180.x

Setliff, A. E., & Courage, M. L. (2011). Background television and infants' allocation of their attention during toy play. *Infancy, 16*(6), 611–639. https://doi.org/10.1111/j.1532-7078.2011.00070.x

Chapter 3:

Davies, A. (2021, October 15). Baby poop 101: A comprehensive guide to newborn and infant poop. *The Bump.* https://www.thebump.com/a/baby-poop

Meek, J. Y., & Noble, L. (2022). Policy statement: Breastfeeding and the use of human milk. *Pediatrics, 150*(1), e2022057988. https://doi.org/10.1542/peds.2022-057988

Moon, R. Y., Carlin, R. F., Hand, I., & Task Force on Sudden Infant Death Syndrome & Committee on Fetus and Newborn. (2022). Evidence base for 2022 updated recommendations for a safe infant sleeping environment to reduce the risk of sleep-related infant deaths. *Pediatrics, 150*(1), e2022057991. https://doi.org/10.1542/peds.2022-057991

Karp, H. (2015). *The happiest baby on the block: Fully revised and updated second edition: The new way to calm crying and help your newborn baby sleep longer.* Bantam.

Nakutavičiūtė, J. (2022, April 17). How to tell if your baby monitor's been hacked—and the steps to take. *NordVPN Blog.* https://nordvpn.com/blog/baby-monitor-hacked/

Richter, D., Krämer, M. D., Tang, N. K., Montgomery-Downs, H. E., & Lemola, S. (2019). Long-term effects of pregnancy and childbirth on sleep satisfaction and duration of first-time and experienced mothers and fathers. *Sleep, 42*(4), zsz015. https://doi.org/10.1093/sleep/zsz015

Wang, A. B. (2018, December 20). "I'm in your baby's room": A hacker took over a baby monitor and broadcast threats, parents say. *The Washington Post.*

Chapter 4:

Archer, C., & Kao, K.-T. (2018). Mother, baby and Facebook makes three: Does social media provide social support for new mothers? *Media International Australia, 168*(1), 122–139. https://doi.org/10.1177/1329878X18783016 SAGE Journals

Babetin, K. (2020). The birth of a mother: A psychological transformation. *Journal of Prenatal & Perinatal Psychology & Health, 34*(5), 410–428. https://www.birthpsychology.com/wp-content/uploads/journal/published_paper/volume-34/issue-5/vwmfBdzE.pdf birthpsychology.com+1

Caruso, C. (2016). Pregnancy causes lasting changes in a woman's brain. *Scientific American.* https://www.scientificamerican.com/article/pregnancy-causes-lasting-changes-in-a-womans-brain/

Fielde, J. M., & Gallagher, L. M. (2008). Building social capital in first-time parents through a group-parenting program: A questionnaire survey. *International Journal of Nursing Studies, 45*(3), 406–417. https://doi.org/10.1016/j.ijnurstu.2006.09.008

Hoekzema, E., Barba-Müller, E., Pozzobon, C., Picado, M., Lucco, F., García-García, D., & Vilarroya, O. (2017). Pregnancy leads to long-lasting changes in human brain structure. *Nature Neuroscience, 20*(2), 287–296. https://doi.org/10.1038/nn.4458

Joseph, N. T., De los Santos, T., & Amaro, L. (2022). Naturalistic social cognitive and emotional reactions to technology-mediated social exposures and cortisol in daily life. *Biological Psychology, 173*, 108402. https://doi.org/10.1016/j.biopsycho.2022.108402

Kirkpatrick, C. E., & Lee, S. (2022). Comparisons to picture-perfect motherhood: How Instagram's idealized portrayals of motherhood affect new mothers' well-being. *Computers in Human Behavior, 137*, 107417. https://doi.org/10.1016/j.chb.2022.107417

Lu, L. (2006). The transition to parenthood: Stress, resources, and gender differences in a Chinese society. *Journal of Community Psychology, 34*(4), 471–488. https://doi.org/10.1002/jcop.20110

Minkin, R., & Horowitz, J. (2023). *Parenting in America today.* Pew Research Center. https://www.pewresearch.org/social-trends/2023/01/24/parenting-in-america-today/

Miller, B. C., & Sollie, D. L. (1980). Normal stresses during the transition to parenthood. *Family Relations, 29*(4), 459–465.

Schoppe-Sullivan, S. J., Yavorsky, J. E., Bartholomew, M. K., et al. (2017). Doing gender online: New mothers' psychological characteristics, Facebook use, and depressive symptoms. *Sex Roles, 76*, 276–289. https://doi.org/10.1007/s11199-016-0640-z

ScienceDaily. (2016). Study shows which new moms post the most on Facebook: Perfectionists and those seeking validation as mothers post more. https://www.sciencedaily.com/releases/2016/05/160524144913.htm

Chapter 5:

American Academy of Pediatrics. (2016). *Media and young minds.* Pediatrics, 138(5), e20162591. https://doi.org/10.1542/peds.2016-2591 Pediatrics Publications+1

American Academy of Pediatrics. (2021). Give your child's eyes a screen-time break: Here's why. *HealthyChildren.org.*

Anderson, D. R., & Pempek, T. A. (2005). Television and very young children. *American Behavioral Scientist, 48*(5), 505–522. https://doi.org/10.1177/0002764204271506

Bergmann, C., Dimitrova, N., Alaslani, K., Almohammadi, A., Alroqi, H., Aussems, S., … Mani, N. (2022). Young children's screen time during the first COVID-19 lockdown in 12 countries. *Scientific Reports, 12*(1), 2015. https://doi.org/10.1038/s41598-022-05840-5

Chen, W., & Adler, J. L. (2019). Assessment of screen exposure in young children, 1997 to 2014. *JAMA Pediatrics, 173*(4), 391–393. https://doi.org/10.1001/jamapediatrics.2018.5546

Christakis, E. (2017). *The importance of being little: What young children really need from grownups.* Penguin.

Christakis, D. A., Zimmerman, F. J., DiGiuseppe, D. L., & McCarty, C. A. (2004). Early television exposure and subsequent attentional problems in children. *Pediatrics, 113*(4), 708–713. https://doi.org/10.1542/peds.113.4.708

Hrdy, S. B. (2009). *Mothers and others: The evolutionary origins of mutual understanding.* Harvard University Press.

Lansbury, J. (2014). *Elevating child care: A guide to respectful parenting.* JLML Press.

Madigan, S., Browne, D., Racine, N., Mori, C., & Tough, S. (2019). Association between screen time and children's performance on a developmental screening test. *JAMA Pediatrics, 173*(3), 244–250. https://doi.org/10.1001/jamapediatrics.2018.5056 JAMA Network+2PMC+2

Myers, L. J., LeWitt, R. B., Gallo, R. E., & Maselli, N. M. (2017). Baby FaceTime: Can toddlers learn from online video chat? *Developmental Science, 20*(4), e12430. https://doi.org/10.1111/desc.12430

Plamondon, A., McArthur, B. A., Eirich, R., Racine, N., McDonald, S., Tough, S., & Madigan, S. (2023). Changes in children's recreational screen time during the COVID-19 pandemic. *JAMA Pediatrics, 177*(6), 635–637. https://doi.org/10.1001/jamapediatrics.2023.0393

Sugiyama, M., Tsuchiya, K. J., Okubo, Y., Rahman, M. S., Uchiyama, S., Harada, T., … Nishimura, T. (2023). Outdoor play as a mitigating factor in the association between screen time for young children and neurodevelopmental outcomes. *JAMA Pediatrics, 177*(3), 303–310. https://doi.org/10.1001/jamapediatrics.2022.5356

Trinh, M. H., Sundaram, R., Robinson, S. L., Lin, T. C., Bell, E. M., Ghassabian, A., & Yeung, E. H. (2020). Association of trajectory and covariates of children's screen media time. *JAMA Pediatrics, 174*(1), 71–78. https://doi.org/10.1001/jamapediatrics.2019.4488

Wolf, C., Wolf, S., Weiss, M., & Nino, G. (2018). Children's environmental health in the digital era: Understanding early screen exposure as a preventable risk factor for obesity and sleep disorders. *Children, 5*(2), 31. https://doi.org/10.3390/children5020031

World Health Organization. (2019). *Guidelines on physical activity, sedentary behaviour and sleep for children under 5 years of age*. World Health Organization.

Chapter 6:

Blumberg, F. C., & Brooks, P. J. (Eds.). (2017). *Cognitive development in digital contexts*. Academic Press.

Gunarti, W., Meilanie, R. S. M., & Marjo, H. K. (2023). The impact of co-viewing on attachment between parents and children. *Jurnal Pendidikan Usia Dini, 17*(1), 31–43. https://journal.unj.ac.id/unj/index.php/jpud/article/view/34862

Hidayati, N., Djoehaeni, H., & Zaman, B. (2022). How parental co-viewing can reduce the adverse effects of gadgets in early children. *ThufuLA: Jurnal Inovasi Pendidikan Guru Raudhatul Athfal, 10*(2), 1–12.

Chapter 7:

Bers, M. U. (2018a). *Coding and computational thinking in early childhood: The impact of ScratchJr in Europe. European Journal of STEM Education, 3*(3), 08. https://doi.org/10.20897/ejsteme/3868

Bers, M. U. (2018b). *Coding as a playground: Programming and computational thinking in the early childhood classroom*. Routledge. https://doi.org/10.4324/9781315398945

Bers, M. U. (2019). Coding as another language: Why computer science in early childhood should not be STEM. In C. Donohue (Ed.), *Exploring key issues in early childhood and technology:*

Evolving perspectives and innovative approaches (Ch. 10). Routledge. https://doi.org/10.4324/9780429457425-11

Blikstein, P. (2018). Maker movement in education: History and prospects. In *Handbook of Technology Education* (pp. 419–437).

Cordes, C., & Miller, E. (2000). *Fool's Gold: A critical look at computers in childhood.* Alliance for Childhood.

Lee, K. T. H., Sullivan, A., & Bers, M. U. (2013). Collaboration by design: Using robotics to foster social interaction in kindergarten. *Computers in the Schools, 30*(3), 271–281. https://doi.org/10.1080/07380569.2013.805676

Oppenheimer, T. (2003). *The flickering mind: Saving education from the false promise of technology.* Random House.

Resnick, M. (2006). Computer as paintbrush: Technology, play, and the creative society. In D. Singer, R. Golinkoff, & K. Hirsh-Pasek (Eds.), *Play = Learning: How play motivates and enhances children's cognitive and social-emotional growth* (pp. 192–208). Oxford University Press.

Robelen, E. W. (2011). STEAM: Experts make case for adding arts to STEM. *Education Week, 31*(13), 8.

Strawhacker, A., & Bers, M. U. (2019). What they learn when they learn coding: Investigating cognitive domains and computer programming knowledge in young children. *Educational Technology Research & Development, 67*(3), 541–575. https://doi.org/10.1007/s11423-018-9622-x

Sullivan, A. (2019). *Breaking the STEM stereotype: Reaching girls in early childhood.* Rowman & Littlefield.

Sullivan, A., & Bers, M. U. (2016). Robotics in the early childhood classroom: Learning outcomes from an 8-week robotics curriculum in pre-kindergarten through second grade. *International Journal of Technology and Design Education, 26*(1), 3–20. https://doi.org/10.1007/s10798-015-9304-5

Sullivan, A., & Bers, M. U. (2018). Investigating the use of robotics to increase girls' interest in engineering during early elementary school. *International Journal of Technology and Design Education.* Advance online publication. https://doi.org/10.1007/s10798-018-9483-y

Wing, J. M. (2006). Computational thinking. *Communications of the ACM, 49*(3), 33–35. https://doi.org/10.1145/1118178.1118215

Yakman, G. (2008). STEAM education: An overview of creating a model of integrative education. In *Pupils' Attitudes Towards Technology (PATT-19) Conference: Research on Technology, Innovation, Design & Engineering Teaching*, Salt Lake City, UT, USA.

Chapter 8:

Barron, B., Martin, C. K., Takeuchi, L., & Fithian, R. (2009). Parents as learning partners in the development of technological fluency. *International Journal of Learning and Media, 1*(2), 55–77. https://doi.org/10.1162/ijlm.2009.0021

Bian, L., Leslie, S. J., & Cimpian, A. (2017). Gender stereotypes about intellectual ability emerge early and influence children's interests. *Science, 355*(6323), 389–391. https://doi.org/10.1126/science.aah6524

Dabney, K. P., Chakraverty, D., & Tai, R. H. (2013). *The association of family influence and initial interest in science.* Science Education, 97(3), 395–409. https://doi.org/10.1002/sce.21060

Dweck, C. S. (2006). *Mindset: The new psychology of success.* Random House.

Google CS Ed Research Group. (2014). Women who choose computer science—what really matters: The critical role of encouragement and exposure (Technical report). Google.

Mannheim, I., & Köttl, H. (2024). Ageism and (successful) digital engagement: A proposed theoretical model. *The Gerontologist, 64*(9), gnae078. https://doi.org/10.1093/geront/gnae078

Master, A., Meltzoff, A. N., & Cheryan, S. (2021). Gender stereotypes about interests start early and cause gender disparities in computer science and engineering. *Proceedings of the National Academy of Sciences, 118*(48), e2100030118. https://doi.org/10.1073/pnas.2100030118

National Girls Collaborative Project. (2024). *State of girls and women in STEM.* https://ngcproject.org/resources/state-girls-and-women-stem

Sullivan, A. A. (2019). *Breaking the STEM stereotype: Reaching girls in early childhood.* Rowman & Littlefield.

The LEGO Group & Geena Davis Institute. (2021). *LEGO ready for girls creativity study.* Geena Davis Institute. https://geenadavisinstitute.org/research/lego-ready-for-girls-creativity-study/

Chapter 9:

Blum-Ross, A., & Livingstone, S. (2016). Families and screen time: Current advice and emerging research. *Media Policy Brief, 17.* London School of Economics and Political Science.

Common Sense Media. (2023). *Common Sense Media: Age-based media reviews for families.* https://www.commonsensemedia.org/

Fairplay. (2023). *Fairplay: Childhood beyond brands.* https://fairplayforkids.org/

Kamenetz, A. (2018). *The art of screen time: How your family can balance digital media and real life.* Hachette UK.

Scratch. (2023). *Scratch.* https://scratch.mit.edu/

Schiller, J. (2023, October 5). Average human attention span by age: 31 statistics. *The Treetop Therapy*. https://thetreetop.com/statistics/average-human-attention-span

Chapter 10:

Bell, T., Alexander, J., Freeman, I., & Grimley, M. (2009). *Computer Science Unplugged: school students doing real computing without computers. New Zealand Journal of Applied Computing and Information Technology, 13*(1), 20–29. https://researchportal.bath.ac.uk/files/214932627/NZJACIT_Unplugged.pdf

Bers, M., Ponte, I., Juelich, K., Viera, A., & Schenker, J. (2002). Teachers as designers: Integrating robotics in early childhood education. In *Information Technology in Childhood Education* (pp. 123–145). AACE.

Cejka, E., Rogers, C., & Portsmore, M. (2006). Kindergarten robotics: Using robotics to motivate math, science, and engineering literacy in elementary school. *International Journal of Engineering Education, 22*(4), 711–726.

Clements, D. H. (1999). The future of educational computing research: The case of computer programming. *Information Technology in Childhood Education Annual, 1*, 147–179.

Elkin, M., Sullivan, A., & Bers, M. U. (2016). Programming with the KIBO robotics kit in preschool classrooms. *Computers in the Schools, 33*(3), 169–186.

Lee, K., Sullivan, A., & Bers, M. U. (2013). Collaboration by design: Using robotics to foster social interaction in kindergarten. *Computers in the Schools, 30*(3), 271–281. https://doi.org/10.1080/07380569.2013.805676

Perlman, R. (1976). *Using computer technology to provide a creative learning environment for preschool children.* MIT, A.I. Laboratory.

Sullivan, A., Elkin, M., & Bers, M. U. (2015). KIBO robot demo: Engaging young children in programming and engineering. In *Proceedings of the 14th International Conference on Interaction Design and Children (IDC '15).* ACM.

Sullivan, A., Kazakoff, E. R., & Bers, M. U. (2013). The wheels on the bot go round and round: Robotics curriculum in pre-kindergarten. *Journal of Information Technology Education: Innovations in Practice, 12*, 203–219.

Sullivan, A., & Strawhacker, A. (2021). Screen-Free STEAM: Low-cost and hands-on approaches to teaching coding and engineering to young children. In *Embedding STEAM in Early Childhood Education and Care* (pp. 87–113). Palgrave Macmillan.

Sullivan, A., & Bers, M. U. (2016). Robotics in the early childhood classroom: Learning outcomes from an 8-week robotics curriculum in pre-kindergarten through second grade.

International Journal of Technology and Design Education, 26(1), 3–20.
https://doi.org/10.1007/s10798-015-9304-5
Wyeth, P. (2008). How young children learn to program with sensor, action, and logic blocks. *The Journal of the Learning Sciences, 17*(4), 517–550.
https://www.jstor.org/stable/pdf/27736742.pdf
Zviel-Girshin, R., Luria, A., & Shaham, C. (2020). Robotics as a tool to enhance technological thinking in early childhood. *Journal of Science Education and Technology, 29*(2), 294–302. https://doi.org/10.1007/s10956-020-09815-x

Chapter 11:

Kardaras, N. (2016). *Glow kids: How screen addiction is hijacking our kids—and how to break the trance* (First St. Martin's Griffin edition). St. Martin's Press.
Martinko, K. J. (2023). *Childhood unplugged: Practical advice to get kids off screens and find balance.* New Society Publishers.
Moise, B. (2024, April 9). Rethink road trips: Tech and tips for traveling with kids. *Verizon News Center.* https://www.verizon.com/about/parenting/family-road-trip-tech-essentials

Conclusion:

DeVon, C. (2023, March 15). 'Parents need to talk about this': The best way to teach your kids about AI, according to an education expert. *CNBC: Next Gen Investing.*
https://www.cnbc.com/2024/03/15/how-parents-can-teach-their-children-to-use-ai-responsibly.html

APPENDIX A. – DIGITAL RESOURCES FOR THE MODERN MAMA

Digital Resources to Support Pregnant and Newly Postpartum Moms		
Category	**Digital Resources to Look For**	**Examples / Creators**
Prenatal Support		
Pregnancy	Pregnancy Tracking Apps	<ul><li>The Bump</li><li>Ovia</li><li>Apps from your physician's office or hospital</li></ul>
Birth Prep and Labor	Birth Prep Classes / Online Education Contraction Timer Apps Podcasts and Videos	<ul><li>Mommy Labor Nurse</li><li>Contraction Master</li><li>iBirth</li><li>Emily Oster</li></ul>
Exercise	Prenatal Yoga, Fitness Videos, Prenatal / Postnatal	<ul><li>Every Mother exercise</li><li>Baby2Body</li></ul>

Digital Resources to Support Pregnant and Newly Postpartum Moms		
Category	**Digital Resources to Look For**	**Examples / Creators**
	Exercise Apps , Diastasis Recti / Core Repair Programs	• Kegel Trainer
Postpartum Support		
Infant Feeding	Breastfeeding, Bottle Feeding, and Pumping	• Milky Mama • Medela Breastfeeding University • Kelly Mom Parenting and Breastfeeding • La Leche League International • One With the Pump • Karrie Lochler
Toddler Feeding	Baby led weaning	• Solid Starts • Baby Led Kitchen
Behavior	Behavior Management, Sensitive Feelings, Sibling Dynamics, Transitions and Habits	• Dr. Becky's "Good Inside" • Janet Lansbury "No Bad Kids"

ABOUT THE AUTHORS

Dr. Amanda Strawhacker and dr. Amanda Sullivan are child development specialists, researchers, educators, and moms who provide expert guidance on educational technology and science, technology, engineering, art, and mathematics (STEAM) education for young children. Affectionately known as "the Amandas" by the children they work with, they bring over a decade of experience in designing and implementing curricula for subjects ranging from robotics and coding to bioengineering in early childhood and elementary settings.

Both hold Master's and Ph.D. degrees in Child Study & Human Development from Tufts University, where they worked with the Developmental Technologies Research Group (now located at Boston College). Together, they focus their collective work on the vital intersection of child development and emerging technologies.

Their mission is to equip parents, caregivers, and educators with evidence-based resources to effectively navigate and utilize technology to foster children's innate curiosity, and promote foundational STEAM learning. Their joint work has resulted in numerous publications, including the book *Playful STEAM Learning in the Early Years: An Educator's Guide to Screen-Free Explorations*, published by Teachers College Press in 2025.